A NEW LIGHT IN PHYSICS

A NEW LIGHT IN PHYSICS

Alejandro Martínez Castillo
Electrical Engineer

Salto
Uruguay

ACKNOWLEDGMENT

To the internet community without which this work would have not been possible.

AMAZON
www.amazon.com

ISBN 10: 1419601202
ISBN 13: 978-1419601200

First version of the book: 2005/January.
This is the 2026/April update.

Things corrected, things changed and new things developed since the first
version...

CONTENT **page**

IMPORTANT NOTE:

> *I'm not infallible (by the way, I make mistakes may be everyday...). Adjustments could be necessary.*

QUOTES:

> *"Sometimes we don't see what we are not looking for..."*

> *"Sometimes we don't see what we don't want to see..."*

INTRODUCTION

WHAT YOU WILL FIND INSIDE:

You will find a new theory about the elementary particles and forces of the Universe.

You will find a new definition of the Electric and Magnetic Forces.

You will find a new theory about how the Electric and Magnetic Forces produced by the elementary particles can interact with those of other elementary particles to form basic particles (photons, electrons, protons, neutrons and neutrinos) and to produce the phenomena of diffraction pattern, photons' absorption, photons' emission, pair annihilation and pair creation.

You will find what light really is. Light and every kind of radiation are made by photons. The photons are particles and here you can see exactly what kind of particles they are and how they exhibit a "wave-like" behavior.

You will find De Broglie relation corrected and with a new physical meaning.

You will find how the mystery of "wave-particle duality" is finally solved.

You will find a new model for the atom and the "subatomic" particles detected experimentally in High Energy Physics.

You will find a new interpretation to the formula $E = mc^2$.

You will find a new interpretation of experiments like those of Kaufmann-Bucherer-Neumann, Davisson-Germer, Young, Hertz, Fizeau, Sagnac, Bertozzi and those experiments about diffraction of photons, electrons diffraction, Doppler Effect, Transversal Doppler Effect.

You will find how communications actually happen due to photons' transmission.

Briefly, a new theory in Physics is proposed in this text. There are modifications made to some current theories and some others are abandoned. The text begins with a correction on the Electric and Magnetic Forces. Those readers who prefer to see first the new theories about photons and the elementary particles of nature can go directly to Chapters Three and Four.

CHAPTER ONE

1.1 BASIS OF THE NEW THEORIES

Here is presented a summary of the main considerations on which the new theories are constructed over.

I) Currently it is assumed that the elementary particles like the electron are "point charges" with no dimension although with a magnetic spin what is no easy to accept by the common sense. Here is proposed a structure for the electron, the photon, the neutrino, the proton and the neutron. It is proposed the existence of a couple of more elementary particles: the *positrin* and the *negatrin*. They generate the known Electric Force, the Magnetic Force, the Gravity Force and a new Ultimate Force (which prevents them from being destroyed in collisions). As we will see all the basic atomic particles are simple combinations of positrins and negatrins.

II) Classically the Electric Field and Force are determined by the Coulomb's law, and the Magnetic Field and Force by the laws of Biot-Savart, Ampere and Lorentz. But these laws have an empirical origin, they were found through measurements in many experiments and there's nothing theoretical that determines they must be that way especially at very high velocities. Here is proposed that is needed a factor of correction in the formulas of the Electric and Magnetic Forces to take into account the behavior of the particles when high velocities are present.

III) Relativity Theory is demonstrated wrong in Appendix A in consequence the following hypothesis are assumed valid:

a) There exist *Rest Frames of Reference* in the Universe. Things that are at rest relative to these frames have no movement. Moving objects have a not zero velocity v_A relative to them and it is their *absolute velocity*.

b) It is considered the Classical *space* and *time* where the space is the Euclidean one where the distance is the Euclidean distance, and the time t is an independent variable that measures the temporal interval between events. Particularly, the addition of velocities is the classical one:

$w = u + v$

c) The velocity of light is not the same in every frame of reference, it follows the vector additive classical law described above:

$ç_2 = ç_1 + u$

where u is the velocity between the frames.

Here and ever in the text we adopt $ç$ as the symbol of light velocity.

$ç$ is not a constant like c.

d) The Emission Theory which proposes that the velocity of light depends on the velocity of the source is then necessary valid (Michelson-Morley experiment).

The final velocity is:

$ç = c + u$

where c is de light constant at which the source emits photons, u is the velocity of the source and vector addition is assumed for the velocities (bold means vector).

The Emission Theory is analyzed in Chapter Eight.

e) As in Classical Physics the Electric and Magnetic Fields are assumed instantaneous at every point of the space and do not propagate at some finite velocity. It is criticized in this text, in Section 7.2, the experiments of Hertz that seems to prove the existence of electromagnetic waves and a finite velocity propagation of the Electric and Magnetic Fields and Forces. An alternative interpretation of the phenomena with photons is presented.

IV) Experimentally are known three main physical behaviors of the basic particles of nature:

a) The "wave-like" behavior of the particles where they verify the De Broglie law.

b) The quantum behavior of the energies of photons.

c) The "subatomic" particles detected experimentally in High Energy Physics.

In this text we are going to describe a new way of understanding them.

CHAPTER TWO

2.1 THE CLASSICAL ELECTRIC AND MAGNETIC FIELDS AND FORCES

In Classical Physics the concept of basic charge exists. The charges are said to generate the Electric and the Magnetic Fields and Forces.

The Electric Field of a point charge is a central one with spherical symmetry that represents an attractive force between opposite charges and a repulsive force between equal charges.

The Electric Field generated by a charge Q is expressed by:

$$E = KQ/r^2 a_r \qquad \text{(bold means vectors)}$$

Where r is the distance to the considered point and a_r is the unity vector in the direction from the charge to the point.

The Electric Force of the charge Q over a charge q is the Coulomb's Force:

$$F_E = qE$$

$$F_E = KQq/r^2 a_r$$

It must be noted that the notion of Electric Field is a mathematical abstraction related to the Electric Force that would eventually act over a particle with unity charge ($q=1$) if it would exist in a specific position of the space.

Classically it is said that the Magnetic Force exists where moving charges appears. The movement must be considered relative to a *Rest Frame of Reference* as described in Appendix A – item B. Moving charges generates a Magnetic Field and an interaction with other moving charges results in a Magnetic Force acting between them.

It is important to consider here curvilinear distributions of charge (following a curvilinear path) and we have:

a) The Magnetic Field generated by some linear distribution of charge is classically expressed by the Biot-Savart law:

$d\boldsymbol{B} = (\mu_0/4\pi)I(d\boldsymbol{l} \times \boldsymbol{a}_r)/r^2$ (bold means vectors)

where $\boldsymbol{a}_r$ is the unity vector in the $\boldsymbol{r}$ direction

To find the complete value of $\boldsymbol{B}$ is necessary to integrate over a path.

It is noted that the notion of Magnetic Field is a mathematical abstraction related to the Magnetic Force that would eventually act over an element of unity charge ($q=1$) if it would exist in a specific position of the space.

b) The force of a field $\boldsymbol{B}$ over a differential current element $\rho d\boldsymbol{l}\boldsymbol{v} = I d\boldsymbol{l}$ where v is the velocity of the charge element in the direction of $d\boldsymbol{l}$ (which is tangential to the element):

$d\boldsymbol{F_B} = \rho d\boldsymbol{l}\boldsymbol{v} \times \boldsymbol{B} = I d\boldsymbol{l} \times \boldsymbol{B}$ where $\boldsymbol{B} = \boldsymbol{B}(r)$

To find the complete value of $\boldsymbol{F_B}$ is necessary to integrate over a path.

If we consider a Magnetic Field $\boldsymbol{B}$ acting over a point charge we have the Lorentz's force expressed as:

$\boldsymbol{F_B} = q\boldsymbol{v} \times \boldsymbol{B}$

where $\boldsymbol{v}$ should be the absolute velocity and $\boldsymbol{B}$ is the Magnetic Field.

NOTE

The classical Electric and Magnetic Fields verify Maxwell Equations.

EXAMPLE

An important example to us is the Magnetic Field generated by a circular ring with radius R. If we consider a frame of reference centered at the center of the ring and the z axis parallel to the ring axis, it can be determined from the equation in a) that the field along the z axis is determined by: [4]

$B(z) = \frac{1}{2}\,\mu_0 I R^2 / (R^2 + z^2)^{3/2}$

If we consider large distances $z >> R$ then:

$B(z) \approx \frac{1}{2}\,\mu_0 I R^2 / z^3$ which is proportional to $1/z^3$

Here we can see that, at small distances, the Magnetic Field can be stronger than the Electric Field which is proportional to the inverse of the square of the distance.

2.2   NEW ELECTRIC AND MAGNETIC FORCES

It is proposed that the Electric and Magnetic Forces formulas must be corrected.

We are going to make a modification on the forces' formulas that will represent the behavior of the basic particles like the electron and the proton at high velocities. It is proposed that the new Electric and Magnetic Forces are directly responsible for that behavior.

When particles travel through Electric and Magnetic Fields, it has been experimentally determined that a factor $s = (1 - v^2/c^2)^{1/2}$ appears, although it is only visible when high velocities are present in the particles. It is well known, for

14

example, that when electrons travel through a strong magnetic field it describes a circular trajectory that verifies the equation:

$$qvB = (m_0/s)(v^2/r)$$

Where:
$$s = (1 - v^2/c^2)^{1/2}$$
c is the light velocity in vacuum emitted by a source at rest, $c \approx 3x10^8$ m/seg.

It is proposed that the factor s is present in the Magnetic Force.

In the example above the equation should be rewritten to:

$$sqvB = m_0v^2/r$$

It is proposed that the factor belongs to the other side of the equation. This gives the same kinematics results but means a different cause to the behavior.

This is the alternative to the proposition of mass variation in the Relativity Theory.

Then, if we denote B^c and F_B^c the Magnetic Field and Force in Classical Physics, it is proposed that the actual Magnetic Field and Force are:

$$\boldsymbol{B} = \boldsymbol{B}^c$$
$$\boldsymbol{F_B} = s_B\boldsymbol{F_B}^c$$

Where is defined:
$$s_B = (1 - v\perp^2/c^2)^{1/2} \text{ for } v\perp \leq c, s_B = 0 \text{ for } v\perp \geq c$$

$v\perp$ is the *perpendicular* to the field **B** component of the *relative* velocity between the *source* of the field and the *particle* where the force is applied.

NOTE:

It must be noted that with the new definition of the Magnetic Force two different velocities are involved in the same formula of the Force. The absolute v_A velocity of the particle relative to a frame at absolute rest, and now the $v\perp$ velocity of the particle relative to the source of the Magnetic Field, present in the s_B new factor.

For the Electric Force F_E, the new theories propose a similar modification, a factor s_E , to take into account the behavior when high relative velocities are present between the source of the field and the particles. If E^c and $F_E{}^c$ are the classical Electric Field and Force then the actual ones should be:

$$\boldsymbol{E} = \boldsymbol{E}^c$$

$$\boldsymbol{F_E} = s_E \boldsymbol{F_E}^c$$

Where is defined:

$s_E = (1 - v_\parallel{}^2/c^2)^{1/2}$ for $v_\parallel \leq c$, $s_E = 0$ for $v_\parallel \geq c$

$v_\parallel$ is the *parallel* to the field $\mathbf{E}$ component of the *relative* velocity between the *source* of the field and the *particle* where the force is applied.

It is interesting to note that the Electric and Magnetic Forces become zero for velocities higher than c. The new theories do not limit the velocity of every object to be less than the value c but as the forces becomes zero at this speed it seems not to be possible to accelerate something to a speed higher than c. As we noticed at the beginning of the text, the new theories determines that the Emission Theory must be valid and so velocities higher than c can be reached if the source of the forces is moving with an absolute velocity $\boldsymbol{u} > 0$.

NOTE

The Electric and Magnetic Fields are the same as in Classical Physics and so verify Maxwell Equations.

The corrections apply to the Electric and Magnetic Forces only.

Next section describes well known experiments that can be successfully explained with the correction in the forces.

2.3   NEW INTERPRETATIONS FOR OLD EXPERIMENTS

I) The strong Magnetic Field experiment

The previous Section 2.2 mentions a well-known experiment with a Magnetic Field. It is the action of a strong uniform Magnetic Force on a beam of electrons. Since the Magnetic Force is always perpendicular to the velocity of the electrons and it is constant a circular path is followed by them.

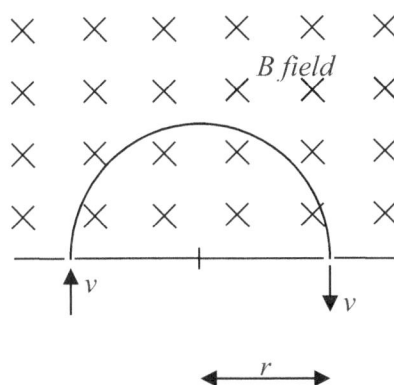

The radius of the path is dependent on the velocity of the electrons. Precise experiments show that with Classical Physics there is a difference between the theoretical calculated radius and the experimentally obtained one. The difference was "perfectly" explained with the factor $s = (1 - v^2/c^2)^{1/2}$ introduced for an electron mass variation. The following equation that relates the Magnetic Force and the Centripetal Force is obtained:

$$qvB = (m_0/s)v^2/r$$

The theory of this text interprets that the factor $s = s_B$ simply belongs to the other side of the equation:

$$s_B q v B = m_0 v^2 / r$$

It is proposed in this text that actually the Magnetic Force is affected by the factor s_B. The same kinematics behavior is predicted with no mass variation needed. The Magnetic Force is:

$$\mathbf{F}_B = s_B q \mathbf{v} \times \mathbf{B}$$

II) Kaufmann-Bucherer-Neumann experiments

We will consider here the experiments performed by Kaufmann, Bucherer and Neumann in the years of the development of Relativity Theory.

In the first Kaufmann's experiment, a beam of electrons traveling in the x direction is simultaneously subjected to the action of a uniform Electric Force producing a deviation in the y direction and a uniform Magnetic Force producing a deviation in the z direction perpendicular to the first. A parallel photographic plate, perpendicular to the x direction, records an impression of the rays and allows a direct measurement of y and z.

With the classical formulas for the E and B forces and the classical kinematics equations a parabola would be expected to be seen in the photography.

But this doesn't happen. The observed curve is different.

The experiment was later improved by other physicists like Bucherer and Neumann.

To agree with the results the proposition made by some physicists some years before, that the mass of the electrons should vary with velocity was considered.

At the end, the Lorentz's factor $k = (1 - v^2/c^2)^{-1/2}$ was associated to the mass.

A new approach is taken here in this text considering that a bad interpretation of the experiments happened and it is proposed that actually the Magnetic Force is affected by the factor $s_B = (1 - v\perp^2/c^2)^{1/2}$ $(v\| = 0, s_E = 1)$.

If the Magnetic Force is affected by this factor the same kinematics' result is obtained while the mass remains constant.

III) Bertozzi experiment

Experiments with strong Electric Fields are performed in electrical accelerators. A new interpretation for the well-known Bertozzi experiment, with the new definition of the Electric Force, is developed here.

The aim of Bertozzi experiment was to verify experimentally the relationship between the velocity of electrons accelerated in an electrical accelerator and their Kinetic Energy.

The classic prediction gives the equation $E_k^c = mv^2/2$ while the relativistic one gives $E_k^r = (1/s - 1)mc^2$ where $s = (1 - v^2/c^2)^{1/2}$.

Bertozzi's results agreed with the relativistic prediction resulting in electrons with velocities reaching asymptotically to the light's velocity c while their acquired energies increase infinitely. This would be in accordance with the relativistic prediction of an infinite increase in the mass of the electrons as they tend to reach the light's velocity c.

Considering that the Electric Energy applied on the electrons by a voltage V is qV, the equation verified in Bertozzi's experiment is:

$$qV = (1/s - 1)mc^2$$

The equation relates the applied voltage V and the velocity v of the electrons.

We will derive here the relation between the Electric Potential U and the final velocity v of the electrons according to the new theory proposed here.

The Electric Force is defined as $F_E = sF_E^c$ where $s = s_E$ and F_E^c is the classic one and so the Electric Force in the electrical accelerator is:

$F_E = sqE^c = sqU/d$

E^c is constant, U is the Electric Potential between the accelerator plates and d their distance.

As the real equation of force is $F = ma$ then:

$sqE^c = ma$

$qE^c = ma/s$

Integrating both sides we have:

$\int_o^d qE^c dx = \int_o^d (ma/s)dx$

It follows:

$\int_o^d qE^c dx = qE^c \int_o^d dx = qE^c d = qU$

$\int_o^d (ma/s)dx = \int_o^t (ma/s)\ vdt = m \int_o^v (v/s)dv = m \int_o^v (v/(1 - v^2/c^2)^{1/2})dv = -mc^2 s|_o^v = -mc2(s - 1) = (1 - s)\ mc^2$

Then, the equation relating the Electric Potential U between the plates and the final velocity v of the electrons is:

$qU = (1 - s)mc^2$

The equation is different from that considered by Bertozzi:

$qV = (1/s - 1)mc^2$

The results of the experiment validate this equation with direct measurement of the voltage V by a voltmeter and a precise measurement of the velocity v of the electrons.

It can be observed on the two equations above that, if it is verified $V = U/s$, both equations are equivalent.

It is argued here that there could be a problem related to the measured "voltage" V in the accelerator plates. There could be a problem with the voltmeter related to its internal functioning.

A voltmeter is a galvanometer connected in series with a high resistance which actually measures an internal current I. A voltage V results applying Ohm's Law $V = RI$.

It must be considered that an electric conductor actually acts as an electrical accelerator of electrons where the Electric Force inside is affected by the factor s for a velocity v reached by the electrons which would depend on the applied Electric Potential only. It is argued here that for the real Electric Potential the law would be affected by the same factor s: $U = sRI$.

Although the relation $V = RI$ remains valid, the real Electric Potential at the terminals of a voltmeter would be related to its internal electric current by: $U = sRI$.

The real Electric Potential U and the measured "voltage" V by a voltmeter would be then related by:

$$U = sV = V/(1 + qV/mc^2)$$

The new theory would demand then some important review in the functioning of the voltmeters and in the way their measurements are applied.

The Electric Energy applied to an electron can be calculated:

$$W_E = \int_0^d F_E dx = \int_0^d sqE^c dx = qE^c \int_0^d s dx = qU(\int_0^d s dx)/d$$

From equation above:

$$s = 1 - qU/(mc^2)$$

It can be calculated:

$$W_E = qU(1 - qU/(2mc^2))$$

2.4   FORCE BETWEEN TWO MAGNETS

A magnet is a piece of material with magnetic properties and it has a north and a south pole. The lines of the Magnetic Field **B** turn around the magnet getting out from the north pole and getting in again by the south pole. The first experiment with magnets was to see that if two equal poles were tried to join together a repulsive force appears and if the opposite poles were placed together an attractive force appears. It is assumed today that inside the magnets, the atoms produce individually equally oriented micro-magnetic-fields. In a permanent magnet many of the micro-magnetic-fields are oriented in the same direction producing a relatively strong macro-magnetic-field. It would be very difficult to determine theoretically the field of a magnet and even more difficult would be to determine the force between two magnets.

Experiments have been made and it seems that the force between them depends on their geometry. In the case of disk magnets a force proportional to the inverse of exponential four in the distance has been found. [3]

F_B is proportional to $1/r^4$

We can see again that, at small distances, the Magnetic Force can be stronger than the Electric Force, which is proportional to the inverse of exponential two in the distance. We'll see the importance of this in the next chapter.

CHAPTER THREE

3.1   PARTICLES AND FORCES

It is considered that *elementary particles* exist in the Universe with laws of their *interaction* and *behavior*.

The possible *interactions* are like attractions and repulsions and are determined by the concept of forces. Four forces are identified in these new theories: the Electric, the Magnetic, the Gravitational and the Ultimate Forces.

All are "action at a distance" forces what cannot be denied. This means that a "Physics System" would exist "running" the Physics Laws on the elementary particles. This leaves us to think in a mathematically based Universe that would "run" in some kind of "Universal Supra-Computer".

The *behavior* of the particles is determined by the laws of Physics, mainly by the Force Law: $F = ma$ demonstrated valid in Appendix A.

The *elementary particles* can then be considered as "sources" of the four forces.

The concept of *particle* can have the following definition:

The elementary particles are geometrical figures in the Space determined by the shape of the origins of the fields of forces.

The Fields are mathematical representation for the Forces. They are considered as related to the Forces that would eventually act over a particle with unity charge ($q=1$) and unity mass ($m=1$) if it would exist in a specific position of the space.

The *charge* is considered as a constant mathematical parameter of the *particles* by which the Electric and Magnetic Fields and Forces can be defined. It is related to the strength of the fields and the shape of their origin. It is not a "material" entity.

The relations between *charge* and Electric and Magnetic Fields and Forces in the new theories are based in the definitions of Classical Physics adding the correction factor *s* to the Forces.

The *mass* is considered a variable mathematical parameter of the particles that appears in the Force Law, the Energy equations and the Gravitational Force. It will be shown directly related to the internal energy of the particles.

The relation between *mass* and the Gravitational Field is also based in the definition of Classical Physics by the Newton's Gravitational Force formula but considering that some corrections are needed as described in Appendix C.

NOTE: The Fields and Forces considered in the new theories are more complex than in Classical Physics.

3.2   THE MOST ELEMENTARY PARTICLES OF THE UNIVERSE

It is proposed that exist a pair of elementary particles in nature and they are called the positrin and the negatrin. One with positive charge and the other with negative charge. The charge has a continuous and curvilinear distribution with the shape of a ring which is the shape of the origin of the Electric and Magnetic Fields of the particles.

The total amount of charge q in each ring is the known value of the electron charge $\underline{e} = 1.6 \times 10^{-19}$ Coulombs. The rings can have angular rotation over their axis and then they will constitute a circular loop of current $I_t = qv_t$ where v_t is the tangential velocity of the charge. In this manner they generate a Magnetic Field and they can exert a Magnetic Force to other elementary particles.

The elementary particles also have associated the peculiar "mass" numerical parameter $\underline{m}$ to which the Force Law $\boldsymbol{F} = \boldsymbol{ma}$ is related:

The Force Law is obeyed by the center of the elementary particles and determines their movement in the Space. The mass $\underline{m}$ represents the inertia of the center of the particles.

The rings of charge have no inertia and are freely to move and rotate around the center just following the laws of the Electric and Magnetic Fields present.

The Gravity Field generated by the elementary particles also depends of the mass parameter $\underline{m}$ and the origin of this field is the center of the rings.

The elementary particles are also the source of a fourth force called the "Ultimate Force". This force is related to a repulsive "Ultimate Field" which must have its origin at a spherical surface containing the rings. It acts as a shield for the particles preventing them from getting too close. At very small distances is stronger than any other force.

3.3   MAGNETIC FORCE BETWEEN TWO ELEMENTARY PARTICLES

We are going to analyze here the magnetic interaction of two particles in a configuration where the rings are parallel and rotating in the same direction with the same v_t tangential velocity.

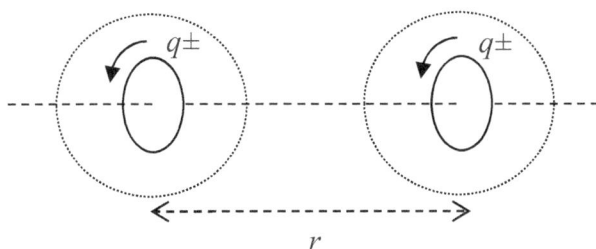

25

The Magnetic Forces between two parallel elementary rings will be solved in steps based on approximations. The approximated solutions are successfully in the goal of explaining the new theory.

NOTE:

In this text is used a non-conventional notation in the equations. The product is considered a dominant operation over the quotient. This means that when a product series appear after the quotient symbol the series is considered belonging to the denominator of the equation (must be computed first).

First step:

We will consider that the two elementary particles are separated by a distance much larger than the radius of their rings. Then, we can apply what was shown in the example of Section 2.1: the field of a ring of current can be approximated at large enough distances by:

$B(z) \approx \frac{1}{2}\mu_0 I_t R^2/z^3$

Then, the field of our ring of the particle that will act on the other ring (the other particle) at a distance r should be:

$B(r)$ proportional to I_t/r^3 with the considered approximation

Second step:

We will consider that the force that one ring exerts on the other should verify the equation in Section 2.1 – item b. But we are not going to compute that. We will follow an approximation approach, taking the geometry of the problem into account and studying the equation:

$dF = \rho v d\boldsymbol{l} \times \boldsymbol{B}$

We can deduce that after the integration the force should be proportional to q, v_t and $B(r)$ ($B(r)$ proportional to I/r^3) and as we assumed that both rings have the same angular rotation (same tangential v_t) we can deduce that:

F_B is proportional to $q^2 v_t^2 / r^3$

We can introduce a constant A to obtain the formula:

$$F_B \approx s_B A q^2 v_t^2 / r^3$$

We will consider small relative velocities $v\perp \approx 0$ then: $s_B \approx 1$.

$$\boxed{F_B \approx A q^2 v_t^2 / r^3}$$

Third step:

We will present now a very important proposition in the new theories.

It is proposed here that in the Magnetism between two elementary rings of charge the effect of Magnetic Induction takes place.

The Magnetic Induction is determined by the quantity of Magnetic Flux generated by one of the rings that passes through the other ring. Some special effects are related on how the Magnetic Induction takes place.

It is proposed here that this phenomenon can be expressed as a very particular variation in the currents of the rings.

It is proposed here as a principle, that the elementary particles exhibit a very particular variation in the v_t velocity of their rings that depends on the distance to the other particle, a special value γ and a factor $f(v_A)$ that is represented by the relationship:

$$\boxed{v_t = \gamma / r^{1/2} f(v_A)}$$

The factor $f(v_A)$ is related to the De Broglie formula and it must be pointed out here that the formula must be corrected:

The Corrected De Broglie equation proposed is: $\lambda = h/mf(v_A)$

The correction with $f(v_A)$ in spite of v_A is relevant at slow velocities only.

$f(v_A)$ should verify:

$f(v_A) \approx v_A$ for high values of v_A

$f(c) = c$

$f(v_A) \approx f(0) \neq 0$ for small values of v_A $(v_A \approx 0)$

This way $\lambda = h/mf(0)$ is not infinite when the particles are at rest $(v_A = 0)$.

NOTE:

The value $f(0)$ is small enough to has been not detected until today in the experiments of electrons, protons and neutrons diffraction.

$f(v_A)$ is not known exactly at the present time but a new version of the Davisson-Germer experiment (see Chapter Six) could, in principle, determine it experimentally.

The Magnetic Force is:

$F_B \approx Aq^2v_t^2/r^3$

Then:

$$\boxed{F_B \approx Aq^2\, \gamma^2/r^4 f^2(v_A) \quad attractive\ or\ repulsive}$$

The determination of the constant $\underline{A}$ still remains to be done.

NOTE:

In the general case of different rings with different absolute velocities:

$\gamma = \gamma_1^{1/2} \gamma_2^{1/2} = (\gamma_1 \gamma_2)^{1/2}$ and $f(v_A) = f(v_{A1})^{1/2} f(v_{A2})^{1/2} = (f(v_{A1}) f(v_{A2}))^{1/2}$

NOTE:

We must consider that magnets are produced with the Magnetic Fields of the particles of their atoms which as we will see in Chapter Four are composed by elementary rings. As mentioned in Section 2.4 the force between two disk magnets with relative large radius has a variation of exponent four in the distance so, F_B proportional to $1/r^4$ is in accordance with the experiments.

NOTE:

We will only note here that the quantum behavior of the photons and electrons can be interpreted in a new way within these new theories by allowing the γ value to vary in a special way. This will be presented with more details in Chapter Five.

3.4   THE EQUILIBRIUM STATES

The elementary particles can interact with other ones through their Electric and Magnetic Forces.

Attractions or repulsions are possible depending on the type of their charges and their Magnetic Fields orientations.

We will analyze here two cases in which two elementary particles reach an equilibrium state.

CASE A:

We will consider now the interaction between particles of opposite type of charge with the same rotation.

The Electric Force at relative large distance is:

$F_E = s_E Kq^2/r^2$ attractive where $s_E \approx 1$ for small relative velocities $v_{\parallel} \approx 0$.

$F_E \approx Kq^2/r^2$

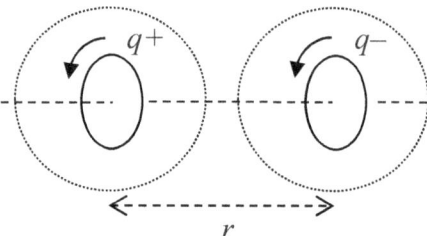

Opposite charges rotating in the same direction generates opposite Magnetic Fields, what means a repulsive force between them.

The Magnetic Force as obtained in Section 3.3 is:

$F_B \approx Aq^2\gamma^2/r^4 f^2(v_A)$ repulsive

Looking at the exponents of distance in each formula we see that, in this case, the repulsive magnetic one is stronger than the attractive electric one at small distances and inversely at large distances. Then, an equilibrium point exists.

We present here now an important statement:

It is proposed that in the case of opposite charged particles the equilibrium distance is $\lambda/2$ where:

_ λ is determined by the Corrected De Broglie formula: $\lambda = h/mf(v_A)$.

_ $f(v_A)$ is a function of the absolute velocity v_A of the particles traveling together and m is the sum of the masses of the two particles involved (m = m_1 + m_2 = $2m_j$).

As it was observed before, the quotient $f(v_A)$ in spite of the v_A of the original equation of De Broglie prevents λ from being infinite when the particles are at rest.

The next graphic made by hand and so with no exact scales and shapes illustrates the situation when particles of opposite type of charge interact to produce the equilibrium state.

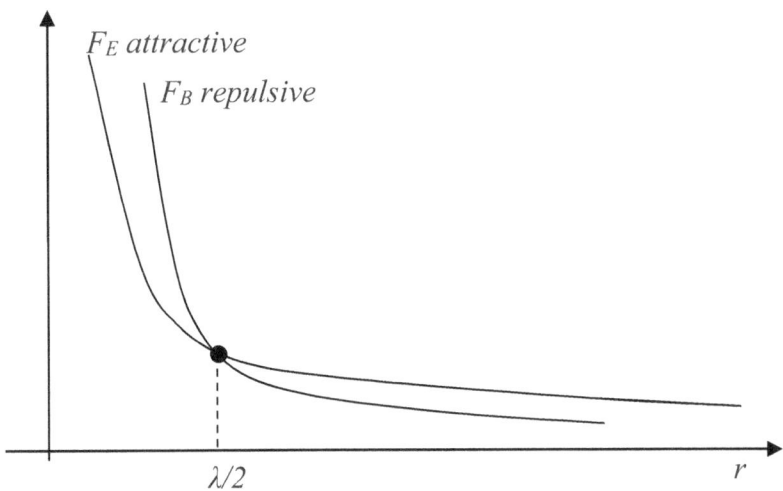

There is a stable equilibrium point because at distances smaller than $\lambda/2$ the repulsive force is stronger than the attractive one and inversely, at distances larger than $\lambda/2$ the attractive is stronger.

In the equilibrium:

$$Kq^2/r^2 \approx Aq^2\gamma^2/r^4f^2(v_A)$$

q cancels out so :

$$K/r^2 \approx A\gamma^2/r^4f^2(v_A)$$

then:

$$r^2 \approx A\,\gamma^2/Kf^2(v_A) \qquad \textit{in the equilibrium}$$

Now we will apply what we have proposed: that the distance in the equilibrium is determined by the Corrected De Broglie formula.

$r^2 = (\lambda/2)^2$ imply that:

$$A\,\gamma^2/K\,f^2(v_A) \approx (h/2mf(v_A))^2$$

The factor $f(v_A)$ cancels in both sides then:

$$m^2 \approx Kh^2/4A\,\gamma^2$$

$$m \approx \tfrac{1}{2}\,(K/A)^{1/2}\,h/\gamma \qquad \textit{in the equilibrium}$$

Is important to remember that the mass <u>m</u> in the Corrected De Broglie formula is the sum of the masses of the two particles: $m = m_1 + m_2 = 2m_j$

The relation above implies that for particles in equilibrium the mass m is independent of the velocity v_A. Then, we will expect that while the couple of particles travel together with an equilibrium relative distance, the mass remains constant independent of their velocity. The exponents of r and $f(v_A)$ in the expression of v_t were chosen to accomplish this condition.

From an external point of view, an intuitive approximation of the lines of the Magnetic Field produced by the two rings of opposite type of charge in equilibrium is pictured by hand in the next figure:

CASE B:

In the case of two particles of the same type of charge rotating in the same direction the Magnetic Fields of both rings have the same orientation and the force between them is attractive.

.

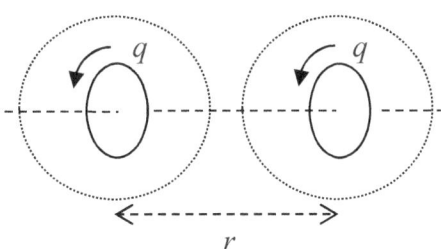

The Magnetic Force between the rings at relative large distances is then:

$F_B \approx Aq^2\gamma^2/r^4 f^2(v_A)$ attractive

And the Electric Force is:

$F_E \approx Kq^2/r^2$ repulsive

Looking at the exponents in distance in each formula we see that the repulsive electric one is weaker than the attractive magnetic one at small distances and inversely at large distances. Then, an equilibrium point exists but it is an unstable one.

$F_E = F_B$ in the equilibrium and as in the case above:

$$Kq^2/r^2 \approx Aq^2\gamma^2/r^4f^2(v_A)$$

$$\boxed{r^2 \approx A\,\gamma^2/Kf^2(v_A) = (\lambda/2)^2 \qquad \textit{in the unstable equilibrium}}$$

The next graphic made by hand and so with no exact scales and shapes illustrates the situation when particles of same type of charge interact to produce the unstable equilibrium state.

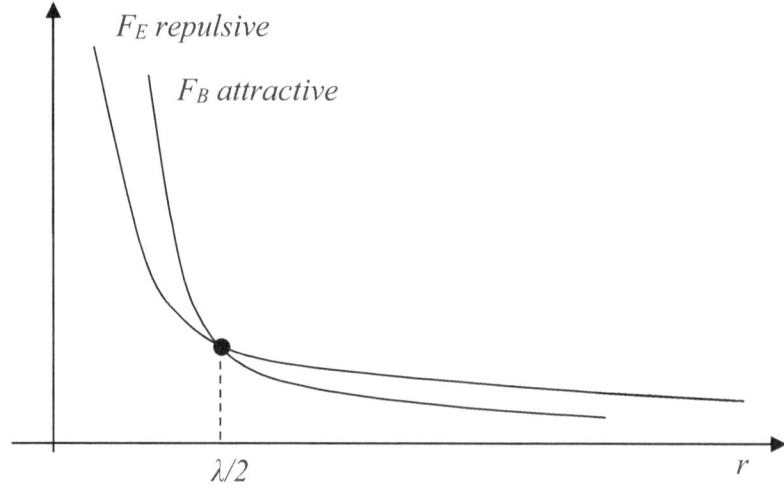

34

From an external point of view, an intuitive approximation of the lines of the Magnetic Field produced by the two rings of equal type of charge in an unstable equilibrium is pictured by hand in the next figure:

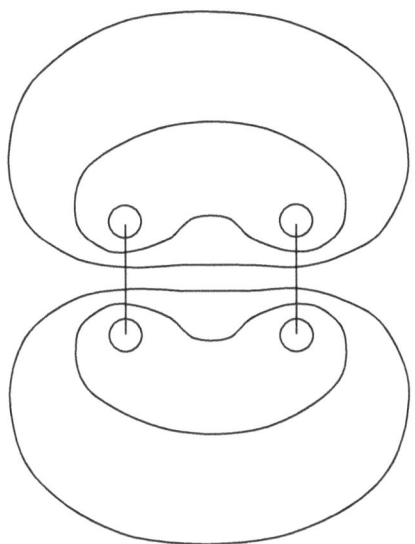

NOTE:

In this section we have proposed a correction in the De Broglie formula and other important consequences result from these new theories:

The De Broglie formula will only apply to the basic atomic particles. It cannot be applied to any object of mass m as Wave Mechanics do.

The value λ has a new meaning: it represents a distance between particles and is not a "wave-length".

There are no "*waves*" associated to the particles. There are no "*waves*" associated to "*matter*". Just a "wave-like" behavior is present.

The Wave Mechanics Theory of De Broglie, Schrodinger, Heisenberg and Dirac is not compatible with these new theories.

The "wave-like" behavior of the particles is described in the next Chapter Four.

CHAPTER FOUR

In this chapter we are going to propose structures for the basic particles: the photon, the electron, the neutrino, the proton and the neutron.

4.1 THE PHOTON

It is proposed that a photon is a pair of a positrin and a negatrin traveling together at light velocity ς and at an equilibrium distance $\lambda/2$ where λ is determined by the Corrected De Broglie formula.

The photon has mass m and verifies the Corrected De Broglie formula (as assumed in Section 3.4 – Case A):

$r = \lambda/2$
$\lambda = h/mf(v_A) \approx h/m\varsigma$

$(\, f(\varsigma) \approx \varsigma, \ f(c) = c \,)$

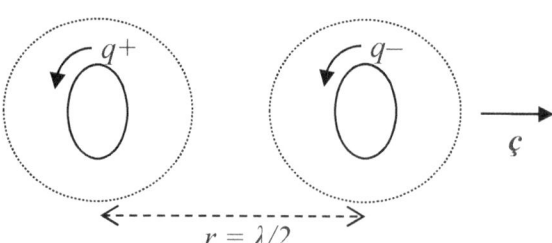

As we mentioned in Section 1.1 the velocity of light, that is the velocity of the photons, is named ς in this text and can vary. Particularly the Emission Theory states that:

$$ç = c + u$$

where u is the velocity of the source of the light and c is the constant velocity at which the source emits the photons.

NOTES:

_ The concept of fields propagating at light velocity sustained by Relativity and Electromagnetic Waves theories are not compatible with this model of the photon because in that case the fields of one ring would not affect the other ring.

They must be considered wrong theories (Chapters One, Two and Seven).

_ Although the particles are traveling at light speed $ç$ the Electric and Magnetic Forces are not affected by the factors s_B and s_E because they are relative factors that involve the velocities $v\perp$ and $v\parallel$ of one particle relative to the fields of the other one and the case is the same as in Section 3.4 – Case A.

_ The E and B fields of the particles that are at rest do not affect the traveling photons because the couple of rings travels at velocity $ç$ relative to that particles ($s_B = s_E = 0$). In the cases of collisions is the Ultimate Force F_U that acts.

The photon has mass and so, for a source at rest ($u = 0$), it has:

Kinetic Energy $E_k = \frac{1}{2}mc^2$

It is known that a photon has the Planck-Energy: $E_{TOTAL} = hv$. This relation is considered to be valid exactly under the assumption of the source at rest only.

If we consider now that $v = c/\lambda$ and that $\lambda = h/mc$, by substitution we have a "Total Energy":

$$E_{TOTAL} = mc^2$$

The difference will be accounted next:

The two rings that compose the photon have Electric and Magnetic fields and there is energy stored in the fields. We will call this energy the Electro-Magnetic Potential Energy P_{EM} of the structure of the photon.

It is proposed that the energy difference in the above relations is due to the Electro-Magnetic Potential Energy P_{EM} that the photon would have in equilibrium at rest:

P_{EM} *at rest* $= \frac{1}{2} \, m_f c^2$

where m_f *is the sum of the masses of the rings (that are equal).*

The Electro-Magnetic Potential Energy P_{EM} at rest (an appropriated zero Electric Potential is defined at Section 4.7) can also be called the "Mass Energy" (*Em*) of the photon. This name remembers us that the value depends in the mass only.

$E_m = \frac{1}{2} \, m_f \, c^2$

Then, actually exists an energy associated with the mass but we have found a different value than that of the Relativity Theory. We found $E = \frac{1}{2}mc^2$. $E = mc^2$ is valid for the photons considering $E = E_{TOTAL} = E_m + E_k$ what means the addition of their "Mass Energy" and Kinetic Energy providing the source of the photons is at absolute rest.

The Total Energy of a particle has a constant component which is the "Mass Energy" and a component variable with velocity which is the Kinetic Energy.

As was mentioned at the beginning, the photon has mass and verifies the Corrected De Broglie formula. There's a correspondence in the mass with the length λ (and so with the frequency v). This implies that the type of photon is determined by its mass while the mass is related to the γ value. Each photon then, has its own γ value. The quantum variation of the special value γ will be treated in details in Section 5.1.

The photon can be called an "electromagnetic particle", a particle with a special electromagnetic structure.

NOTE:

One of the most celebrated experiments by Relativity Theory is the experiment that proved the existence of a curvature of the direction of the light when passing very near to the sun. It is proposed here that the curvature happens simply because the photons have mass ($h\upsilon = mc^2$) and so they are attracted by the Gravitational Force. Calculations for single photons have already been done in the past and showed discrepancy with experimental data but now new calculations for photons traveling in arrangements of trains of photons as proposed in the next Section 4.2 must be done.

4.2   THE PHOTONS' DIFFRACTION

We will see now how this particles exhibit the diffraction behavior.

It is proposed that photons tend to be aligned to form trains of photons:

Linear train of photons

Each photon links to the next by the same equilibrium phenomenon described for the positrin and negatrin that constitutes a single photon. A positrin is always followed by a negatrin. The electric and magnetic forces maintains a strong link.

We will see now how these trains of photons satisfy the Huygens phenomenon. We must consider two trains separated by a distance *d* and emitted simultaneously in a certain angle θ such that they join at a certain "large" distance.

The next schema describes the phenomenon:

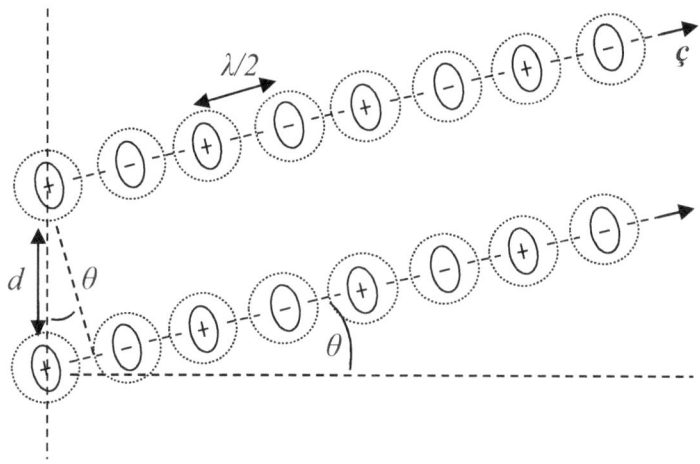

It can be observed that depending on the angle θ the trains will join at a certain large distance with different phases.

In certain angles they will join in phase what means that charges of the same sign of each train are confronted (this happens for example at $\theta = 0$). In this case the repulsive electric force will maintain them apart and they behave separately as ordinary beams of light. They can be reflected or exhibit photons' absorption phenomena and can be detected.

In other angles the trains will join out of phase what means that opposite charges of each train are confronted. In this case there is an attractive force between the trains.

It is proposed that when the trains join out of phase they form a compact array of a couple of trains that cannot be separated and can pass through common materials (between the atoms) or bounce back in a way that there's no photons' absorption phenomena and they are not detected.

40

This is the case that corresponds to the called "destructive interference".

We can deduce from the figure that:

When $sin\theta = n\lambda/d$ $\qquad\qquad$ $n=0,1,2,3...$

"constructive interference" exists.

When $sin\theta = (n + \frac{1}{2})\lambda/d$ $\qquad\quad$ $n=0,1,2,3...$

"destructive interference" exists.

Note that if a beam of light passes through a fine slit some trains interact with the borders and are deviated in all directions and the same phenomenon applies.

4.3   THE PHOTONS' REFRACTION

The refraction is the physical phenomenon that appears when the light passes through the surface border of two media with different light average velocity of propagation. The light obeys the Snell law of refraction that determines the angle at which the light is refracted according to the light velocities in each medium:

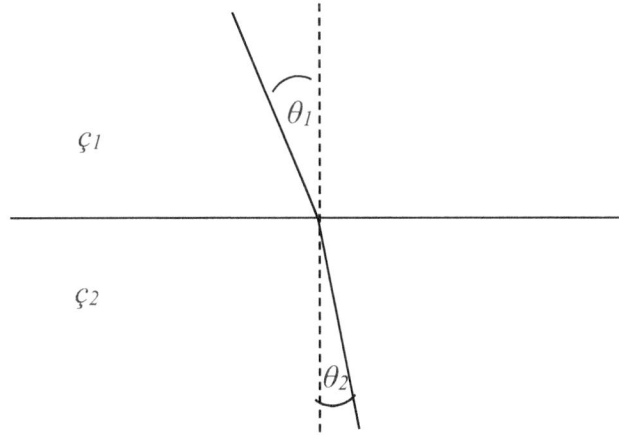

The Snell law states:

$$n_1 sin\theta_1 = n_2 sin\theta_2$$

where: $n_1 = c/ç_1$ and $n_2 = c/ç_2$ are the refraction index of each medium

We will give an explanation of the phenomenon based on the concept of trains of photons.

In general, light is emitted as a three-dimensional arrangement of trains of photons but a two dimensional array is enough to explain the phenomenon because of its symmetry.

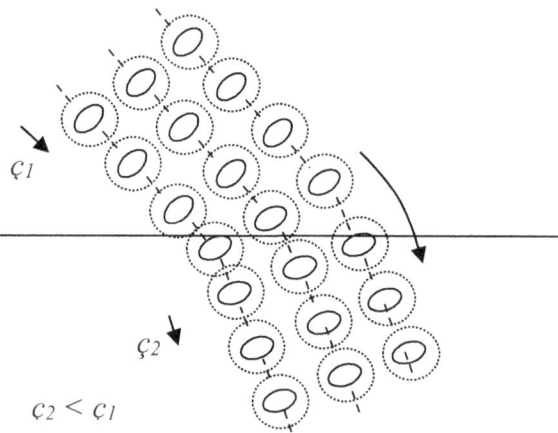

All the photons maintain links with the nearer photons such that the array is compact and it is maintained aggregated while entering the second medium. If the second medium has a slower average velocity the beam is forced to brake on one side first than the other. The figure shows that, if $ç_2 < ç_1$ the left part of the array is comprised because the photons of this side are braked first (by atoms of the medium) than those of the right side. Is intuitive that this makes the array rotate in the border region and this is what produces a new angle of propagation of the array in the second medium.

A prism separate white light in the rainbow spectrum. This could be explained considering a large beam as a large train composed of short sub-trains with different color of photons each (this is in accordance with Newton's experiment of the disk with all colors which when rotating at enough velocity it is seen as a white disk). The refraction angle is dependent on the color of the incident photons and they will refract in different angles each at their time.

4.4   THE ELECTRON

It is proposed that an electron is formed by three elementary particles, one positive and two negative, disposed in a linear configuration such that the positive one is in the center and the negatives at the extremes.

In this manner each negative particle "sees" a positive net charge because the positive one is nearer (the Electric Field of the nearer positive one is stronger). Then, the same phenomenon of magnetic attraction between opposite charged particles applies.

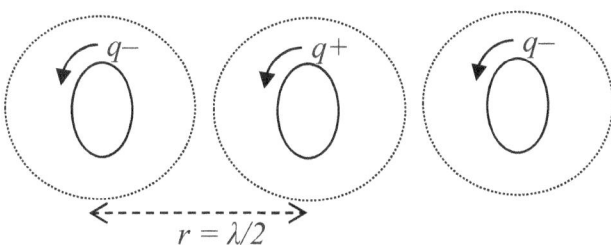

It is proposed here that the formulas that are valid for the equilibrium of the photons are valid for the equilibrium between the tree particles of the electron.

The three particles are in equilibrium at a distance $\lambda/2$ and an Electromagnetic Potential Energy P_{EM} will exist that constitutes its Mass Energy:

$P_{EM} = E_m = \frac{1}{2}mc^2$

where:

m is the total mass of the electron

$m = m_1 + m_2 + m_3 = 3m_j$

The electron verifies the Corrected De Broglie formula:

43

$r = \lambda/2$

$\lambda = h/mf(v_A)$

A free electron has a well defined mass. Experimentally there is no electron mass variation detected nowadays. Then, we must assume that for free electrons γ has a fixed value and we can call it γ_e.

A positron (the anti-electron) has the same structure of the electron but positrins and negatrins are interchanged. It is a positrin-negatrin-positrin sequence.

4.5   THE ELECTRONS' DIFFRACTION

The electron verifies the Corrected De Broglie formula:

$r = \lambda/2, \quad \lambda = h/mf(v_A)$

Where m is the total mass of the conjunct: $m = m_1+m_2+m_3 = 3m_j$

We will see now how the explanation on how the photons exhibit the "wave-like" behavior of diffraction applies to trains of electrons.

It is proposed that electrons tend to be aligned to form trains of electrons.

The trains are formed by the interaction of the negatrins at the extremes as described in Section 3.4 – Case B but not reaching that equilibrium.

It can be observed now that if the electrons establish equilibrium states at a distance λ they can exhibit the same diffraction behavior than the photons.

It is proposed that trains of electrons are formed by equilibrium states between electrons of same magnetic orientation with the negatrins in the extremes staying at a distance $\lambda = h/mf(v_A)$.

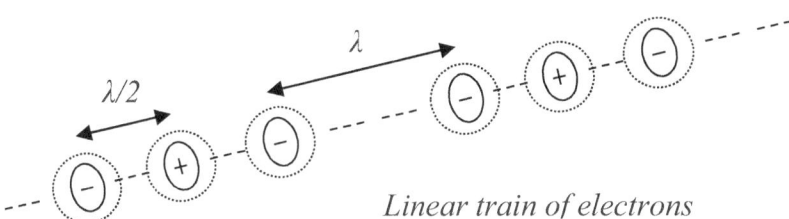

Linear train of electrons

We see that there are two kinds of equilibrium at the same time: the equilibrium between rings with opposite charge at a distance $\lambda/2$ and the equilibrium between rings with same charge at a distance λ.

It can be verified that the train of electrons present the same Huygens behavior as the train of photons (Section 4.2).

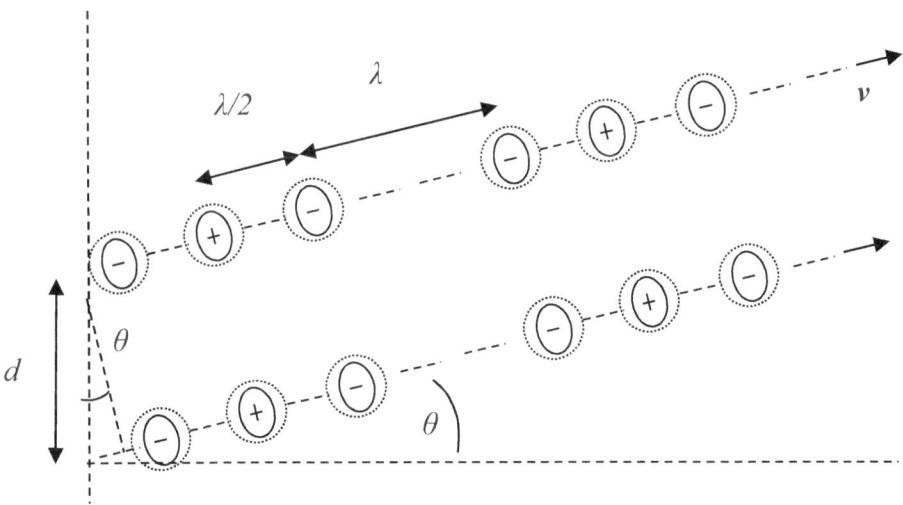

Then, we have a very similar figure of that in Section 4.2.

It can be observed that depending on the angle θ the trains will join at a certain large distance with different phases.

In certain angles they will join in phase what means that charges of the same sign of each train are confronted (this happens for example at $\theta = 0$). In this case the repulsive electric force will maintain them apart and they behave separately as ordinary beams of electrons. The electrons then can be detected by electrons' detectors.

In other angles the trains will join out of phase what means that opposite charges of each train are confronted. In this case there is an attractive force between the trains.

It is proposed that when the trains join out of phase they form a compact array of a couple of trains more difficult to separate which pass through common materials (between atoms) or bounce back in a way that they are not detected.

When $\sin \theta = n\lambda/d$ $n=0,1,2,3...$

"constructive interference" exists: the electrons are captured by the electron detector in these angles.

When $\sin \theta = (n + \frac{1}{2})\lambda/d$ $n=0,1,2,3...$

"destructive interference" exists: the two trains forms a very compacted array of a couple of trains that pass through the electron detector (between its atoms) or bounce back without being detected.

Note that if a beam of electrons passes through a fine slit some trains interact with the borders and are deviated in all directions and the same phenomenon applies.

4.6   ABOUT THE DOUBLE-SLIT EXPERIMENT

The Young's double-slit experiment uses a double slit separated by a distance about some multiples of the "wave-length" of the particles it emits.

Interference patterns of the particles are observed. The peculiarity of the experiment is that it is designed for the emission of individual particles at a time.

The experiment fails in the assumption that individual particles are emitted at a time.

For both electrons and photons a process that decreases the intensity of the beams until discrete events are observed is used. It is currently assumed that this can only happen with individual particles.

It is considered for example that the number of electrons emitted is controlled by controlling the temperature of the heated cathode filament.

But these processes do not guarantee the emission of individual particles.

It is proposed here that in the double-slit experiment actually a burst of parallel trains of particles is emitted at a time and the same phenomena as the diffraction of photons and electrons happen in the same way as described in Sections 4.2 and 4.5.

This possibility has not been considered before just because the concept of trains of particles didn't exist until now.

NOTE:

It must be noted that with the concept of trains of particles the two slit experiment is equivalent to an experiment with one slit where the particles interact with the borders.

NOTE:

Lighting the double slit experiment the diffraction pattern disappear because the lighting particles collide with the trains going to the slit breaking their structure and their particles behaves as isolated individual ones.

The experiment could be improved. Mobile PIN diodes detectors before and after the slits could precisely detect the number of particles present in each discrete event. The experiment would be done first without the diodes and repeated after moving the diodes to the slits.

4.7   THE NEUTRINO

It is proposed that the neutrino is a pair of a positrin and a negatrin rotating in opposite directions.

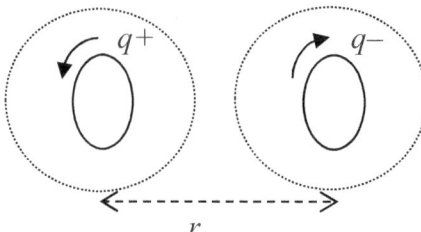

In this case both the Electric and the Magnetic Fields are attractive and the two elementary particles get very close separated by their Ultimate Force.

In this configuration the Corrected De Broglie formula $\lambda = h/mf(v_A)$ does not apply and the composed neutrino particle would remain with very small Internal Energy $E_m = P_{EM} = mc^2/2$ and so very small mass.

It is considered in the new theories that neutrinos exist everywhere and in high amount as they are strictly necessary for the photons' absorption and emission phenomena to happen at any place of the Space. The neutrinos are essential particles that participate in the photons' absorption and emission processes as explained in Section 5.3.

NOTE:

Neutrinos can participate in the experimental interactions between the basic particles at the high energy accelerators as considered in Section 4.11.

NOTE:

Neutrinos are present in the electron and positron pair's creation and annihilation processes as described in Section 5.2.

4.8   PROTONS AND NEUTRONS

Protons and neutrons present diffraction patterns that obey the De Broglie formula. This suggests us that probably protons and neutrons could be made of simple structures like that for the electron.

The elementary particles, the positrin and the negatrin, are so successfully in the representation of the photon, the electron and as we will see in following chapters, in the physical behaviors of them (creation/annihilation of pairs and the photons' absorption and emission), that it is natural to think that they can be the elementary particles with which protons and neutrons can be made of. It is natural to think that they will have structures coherent with the new theories. Particularly it is assumed that they are composed by rings and that these rings can rotate to produce currents that produce Magnetic Fields that can interact with the Magnetic Field of other particles.

It is proposed that the proton is composed by the same structure as the positron. It is a sequence of a positrin a negatrin and a positrin again with the difference in the γ value. A γ_P value with $\gamma_P \ll \gamma_e$ that determines a mass $m_P \gg m_e$.

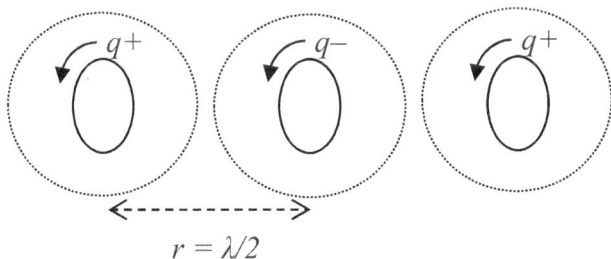

A similar structure is conceived for the neutron:

It is proposed that the neutron has a structure made with four elementary rings, two positrins and two negatrins in an alternating sequence.

49

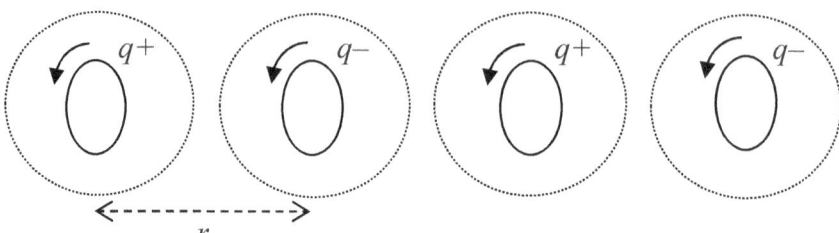

The free neutron (outside the nucleus of an atom) has a natural "decay" into a proton and so its structure is unstable.

This can be explained if we consider that the rings of the neutron have the same γ value than that of the rings of the proton (γ_P) but their v_t is smaller such that the total mass of the four rings in the neutron is quite the same as that of the proton. This configuration is not in equilibrium.

The r distance between the rings of a neutron inside an atom is different from the $\lambda/2$ of a proton but when isolated it verifies the De Broglie relation for the same λ of the proton. This could be because when isolated the neutron is unstable and in process of decaying in a proton.

Currently the *neutron beta decay* is thought as:

$N \rightarrow P + e + antineutrino$

In the new theory it is thought as:

$N + neutrino \rightarrow P + e$

In terms of the structures: *pnpn + pn → pnp + npn*

The neutron interacts with a neutrino to produce a proton and an electron, actually not decay. It must be considered that neutrinos exist everywhere and in high amount.

In Section 5.1 it is presented the possible sets of variation of the special value γ and they determine why only the basic particles (protons, electrons, photons, neutrinos and their "anti-particles" can be stable particles.

In nature there is no other basic stable particle outside the nucleus of the atoms than photons and electrons and their anti-particles so, no stable structures with more than three rings are expected. This, remains to be demonstrated may be with computational simulations of the behavior of the particles.

4.9   NUCLEUS' STRUCTURE

The new theory proposes special structures with the elementary positrins and negatrins for the proton and the neutron. They are presented in Section 4.8.

Here it is described how is the structure of the nucleus of an atom that maintains protons and neutrons together at the very small nuclear dimension.

In the new theory it is proposed that the "Strong Nuclear Force" that binds protons and neutrons together in the nucleus of an atom is the Magnetic Force existent between rings with equal type of charge. If the particles have appropriated magnetic orientation the force is attractive and it can reach an equilibrium state with the repulsive Ultimate Force of the particles at some very small distance.

It must be considered that in the creation of an atom its protons and neutrons are approximated together at such distance.

This Magnetic Force is a very good candidate for the "Strong Nuclear Force" since it has exponent four in the distance.

With the exception of Hydrogen, atoms have many protons and neutrons in their nucleus and structures with many protons and neutrons in equilibrium exist.

It is proposed that all atoms have their protons and neutrons establishing equilibriums at the center of the nucleus with the other protons and neutrons.

They all accommodate in equilibrium states between positive extremes to accomplish the structure of the nucleus. Their other extremes spread out.

The atoms could, in principle, be neutralized by electrons situated near the outer extremes of protons if each proton has one electron associated to it. The electrons also establish equilibrium with associated protons but between the Magnetic and the Electric forces of the rings with opposite type of charge. Section 5.3 describes how discrete equilibrium states between protons and electrons are possible and be in accordance with the "quantum behavior" of atoms in their characteristic photons' emission and absorption spectrums.

The following image illustrates the structure of a neutral helium atom:

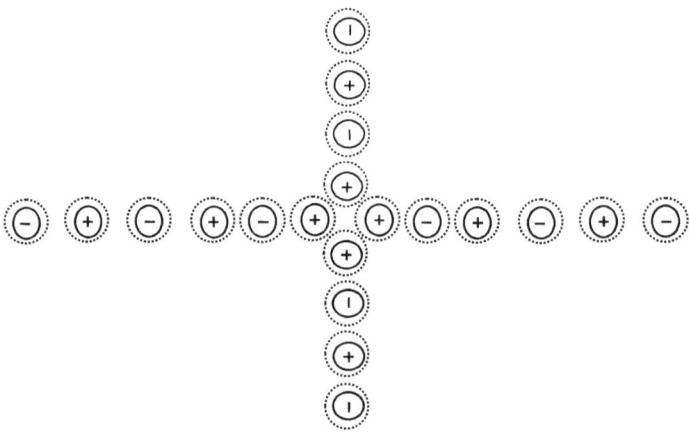

Helium atom

NOTE:
The image does not represent real dimensions of the rings and relative distances. The dimensions of the rings are not known yet.

NOTE:
The structures of the basic particles and that of the atom is based on equilibrium states affected by the factor $f(v_A)$. This implies that the structure of the atoms is velocity dependent.

The structure of "matter" is then velocity dependent (not the mass which remains constant as pointed out in Section 3.4 – Case A).

4.10   THE "SPIN"

Experimentally it is found that the basic particles have a "small" and stable Magnetic Field that is called *Magnetic Moment*.

The currently called *"spin"* of the basic particles like the proton and the electron is directly related to the *Magnetic Moment* measured in each particle.

In the new theories presented, the photon and the electron are composed by rings of currents with Electric and Magnetic Forces in a configuration with equilibrium, so it is very well justified that their Magnetic Field is stable.

We will see now why the *Magnetic Moment* is small:

It was proposed that the basic particles like the photon, the electron and the neutrino are composed by rings of currents that exhibit a repulsive Magnetic Force. This means opposite Magnetic Fields.

A simplified diagram of the lines of the Magnetic Field in the case of photons can be seen in the figure:

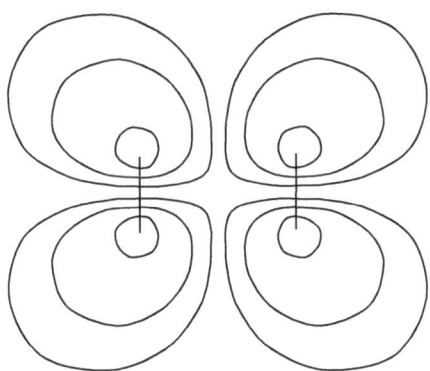

And a diagram for electrons and protons can be seen in the next figure:

The problem that arises here is resumed in this question: What is the average Magnetic Field seen from a relative large distance?

I made a simple experiment with two small disks magnets from audio speakers: If we pass one of them near another relatively strong magnet we can feel the attractive or repulsive forces (depending on the relative position of them) that act between them. But when the two small disks are brought together with opposite magnetic orientations, we need a force to maintain them together and quite no force is felt while passing them around the external magnet even very near of it. The "external" field seems to vanish, the "internal" field continues relatively strong.

What this experiment show is that even if the "external" field seems to be very small, "internally" a strong field can be present. This can be understood if we admit that quite all of both of the Magnetic Fields are confined to a small region around the rings. "Outside" (larger distances) only the weak lines remain.

And this explains why the *Magnetic Moment* is measured experimentally as a very weak Magnetic Field while strong fields are needed to join the positrin and negatrin rings.

4.11   EXPERIMENTALLY DETECTED PARTICLES

Many "strange" particles have been detected experimentally in high energy collisions between protons, neutrons, electrons and atoms.

They are unstable and have a very short lifetime (in the order of microseconds) after which they decay into other particles.

We will see here how those particles could be explained by the new theories.

It is considered here that when elementary particles are close enough and with the same Magnetic Field orientation, the Magnetic Induction as described in Section 3.3 takes place transferring Magnetic Energy and so the called electromagnetic P_{EM} energy from one particle to the other.

The P_{EM} energy is the Electromagnetic Potential Energy stored in the rings as introduced in Section 4.1. It is the *internal energy* of the particles. It is directly related to their variable *mass* parameter m by the relation $P_{EM} = \frac{1}{2}mc^2$ and called Mass Energy E_m in this new theories. Furthermore, the *mass* m can be determined by the relation $m = 2E_m/c^2$ where $E_m = P_{EM}$.

An important feature present in the interactions is that they can have the participation of surrounding neutrinos.

Let we see some possible unstable particles that could be produced in two possible cases that, under certain conditions, a neutron interact with neutrinos. It is considered the neutron with negligible Kinetic Energy (at rest) and so its energy composed by its Mass Energy $E_m = \frac{1}{2}mc^2$.

1)   A neutron N interact with a neutrino v:

In terms of rings: *pnpn + pn → pnp + npn*

Due to Magnetic Induction half of the mass and Mass Energy of the Neutron is transferred to the neutrino and so a pair of a *"half-proton"* and a *"half-antiproton"*, proton-like particles with half the mass and Mass Energy of the proton, can be produced:

$N + \upsilon \rightarrow$ *half-P⁺* + *half-P⁻*

These "*half-P⁺*" *and* "*half-P⁻*" hypothetical particles well correspond to the known "*kaons*" particles K^+ and K^-.

Further, if the structure of the *half-P* particles is broken into their individual rings someway, we obtain new particles with (1/2)/3 = 1/6 the mass and Mass Energy of the original proton.

These new particles well correspond to the known "*pions*" particles Π^+ and Π^-.

2) A neutron N interacts with two neutrinos υ:

In terms of rings: *pnpn + pn + pn → pnpn + pnpn*

Due to Magnetic Induction half of the mass and Mass Energy of the Neutron is transferred to the neutrinos and two "*half-neutrons*", neutron-like particles with half the Mass Energy of the neutron, can be produced:

$N + \upsilon + \upsilon \rightarrow$ *half-N* + *half-N*

This "*half*-N" hypothetical particle well corresponds to the known "*kaon*" particle K^0.

Further, if the structure of the *half-N* particles is broken into their individual rings someway, we obtain new particles with (1/2)/4 = 1/8 the mass and Mass Energy of the original neutron.

These new particles well correspond to the known "*muons*" particles μ^+ and μ^-.

May be other possible interactions can produce results similar to those described above. There can be other ways to produce the same particles.

Other kind of interactions will produce the remaining of the experimentally detected particles. This subject needs further development.

Inexactitudes in the values of the Mass Energy of the produced particles are expected since conversions between Kinetic Energy and Mass Energy of the particles can be present.

The unstable and short life-time characteristic of the produced particles is well explained with this new approach since the γ values of the rings of the particles can only vary in allowed intervals as described in the next Section 5.1 and do not verify the equilibrium conditions described in Section 3.4 – Case A. The produced particles can interact again with other particles to produce yet other kind of particles.

The principles of conservation of energy and the linear and angular momentum are verified in the described interactions. The conservation of the currently named "spin" of the particles could be not conserved and it is because that "spin" is not directly related to the angular momentum as shown in Section 4.10.

NOTE:

It must be reminded here that particles usually travel in arrangements of trains of particles what can leave to confusions like that described in Section 4.6. There is pointed out that in the double-slit experiment actually arrangements of trains of particles are emitted in each "single event" in spite of individual ones and that this is the main cause for the experiment to hasn't been explained successfully until now.

Care must be taken then, when considering the energy of particles in high energy experiments, to not confuse the energy of arrangements of trains of particles with the energy of individual particles.

CHAPTER FIVE

The most typical "quantum" phenomena are the photons' emission and absorption of photons by the atoms and the creation/annihilation of pairs of electrons and positrons. We will interpret them with the new theories.

5.1   THE SPECIAL VALUE γ

We have already presented the particular behavior of the relative current $I_t = qv_t$ of the rings in Chapter Three:

In Section 3.3 was proposed that the elementary particles exhibit a variation in the_angular velocity of the rings that depends on the distance to the other particle and the special value γ that is represented by the relationship:

$$v_t = \gamma/r^{1/2}f(v_A)$$

As we mentioned there, v_t is the tangential velocity of the charge in a ring.

The value γ is "special" in the meaning that it has a special variation.

There are three sets of variation of the γ value with different characteristics:

1) The *"proton set"*:

The γ value in this set determines the existence of protons and neutrons and can be called γ_P. It is characterized for being able to have a *discrete* variation which determines the quantum spectra of electromagnetic radiation when absorbed or emitted by atoms. This variation is shown in Section 5.4 where it is presented the possible values according to the Hydrogen spectrum.

2) The *"electron set"*:

It is characterized by a *unique* constant value γ_e which determines the existence of the electron and the positron. This value is much bigger than the proton's value determining a much smaller mass for the electron in relation to the proton's mass.

3) The *"photons set"*:

It is characterized for being able to have a *continuous* variation in a wide interval from some minimum value (much bigger than the electron value) to some maximum value which are yet unknown and need to be determined. It determines the existence of photons.

The behavior of photons is well described in Section 5.3 where the photons' absorption and emission phenomena are analyzed. In this section we can understand the basic process by which electromagnetic radiation is absorbed or emitted by material objects.

The γ value can vary but only between the allowed values in the sets described above.

The variations in γ can only happen in interactions of particles where energy transferring takes place. A new γ determines a new stable internal equilibrium of a particle with a new internal Mass Energy E_m and so new mass m.

5.2   PAIR CREATION AND ANNIHILATION

The electron and the positron are made by three elementary particles and the photons by two then, in a pair of positron and electron creation/annihilation must be involved two photons and a third pair of elementary particles.

It is proposed that theoretically a third particle exists in a creation or annihilation processes and that it is a neutrino.

I) The annihilation process

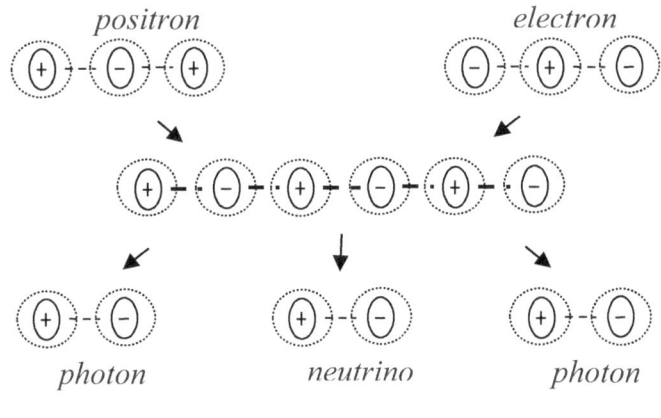

positron　　　　　　　*electron*

photon　　　*neutrino*　　*photon*

Annihilation

In the annihilation process we have:

positron and electron rotating in opposite directions with:

    masses: m_e

    Mass Energy = $E_m = \frac{1}{2} m_e c^2$

    Electromagnetic Potential Energy of the pair: $P_{EM} > m_e c^2$

photons with:

    Total Energy = $E_{TOTAL} = m_f c^2$

    masses: m_f

neutrino with:

    small Kinetic Energy

    small mass

It must be considered that the electron and the positron with opposite rotations are attracted by both the Electric Force and the Magnetic force. They interact in a similar way as two separated rings of opposite charge and rotation tending to get very close and to reach an equilibrium state with their Ultimate Force F_U.

An Electro-Magnetic Potential Energy P_{EM} in them exists. The Electro-Magnetic Potential Energy is the sum of the Electric Potential and the Magnetic Potential Energies.

Note that the particles themselves are in an "inner" equilibrium and have also an "inner" Electro-Magnetic Potential Energy which we call the "Mass Energy".

It is proposed that the Electromagnetic Potential Energy P_{EM} of a positron and an electron with opposite rotations and separated at a distance equal to half of their De Broglie λ is:

$$P_{EM} = \tfrac{1}{2}(m_e + m_p)c^2 = m_e c^2$$

Before the creation of the photons the situation can be thought as an unstable intermediate structure of six rings and, in some way not totally understood yet, one pair of positrin and negatrin becomes a neutrino delivering its Mass Energy to the two pairs of remaining rings that becomes two photons traveling in opposite directions. A quantity equal to $\tfrac{1}{2}(m_e + m_p)c^2 = m_e c^2$ of energy of the pair positron-electron is converted to Kinetic Energy for the photons.

We will consider the situation where the electron and the positron are at some distance $d > λ/2$ and so they have a $P_{EM} > m_e c^2$ expressed as $P_{EM} = m_e c^2 + \Delta E$.

The electron and the positron attract themselves. Both the Electric Force and the Magnetic Force are attractive. There's a natural tendency for the annihilation process to take place.

The energy balance in the annihilation process can be calculated.

Initially, for the electron and the positron we have:

$E_m = \frac{1}{2} m_e c^2$ for each one

$P_{EM} = m_e c^2 + \Delta E$ for the pair

The energy converted for the photons is

$E_{TOTAL} = 2E_m + m_e c^2 = 2m_e c^2$

Finally, for the photons created we have:

$E_m = \frac{1}{2} m_f c^2$ for each one.
$E_k = \frac{1}{2} m_f c^2$ for each one
$E_{TOTAL} = E_m + E_k = m_f c^2 + m_f c^2 = 2m_f c^2$

And a neutrino will be produced with small mass and the remaining ΔE Kinetic Energy just to validate the principle of conservation of energy in the process.

It must be:
$2m_e c^2 = 2m_f c^2$
We conclude that: $m_f = m_e$

The photons created have a mass equal to the mass of the electron or the positron. They will also have the same λ.

II) The creation process

We will consider now the symmetric process: two colliding photons and a little neutrino with just the necessary energies to produce the creation process:

photons with same orientation, opposite rotations and energy: $m_f c^2 = m_e c^2$

neutrino with same orientation and energy: $E_m + E_k = \Delta E$

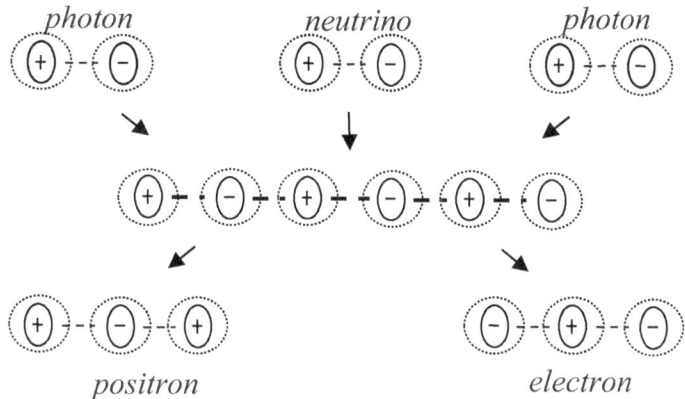

Creation process

 The energy balance in this case is also completely symmetric to the annihilation case.

Total Energy of the two photons and the neutrino:

$E_{TOTAL} = 2m_f c^2 + \Delta E > 2m_e c^2$

The energy for the electron and positron pair with opposite rotations is:

$E_{TOTAL} = E_m + P_{EM} + E_k = m_e c^2 + m_e c^2 + \Delta E$

 It´s important to observe that half of the energy of the photons is used in this creation process to create the positron and the electron but also more energy is needed for the Electromagnetic Potential Energy P_{EM} and the relative Kinetic Energy that exist when they are separated apart. The other half of the energy and the neutrino energy are converted to them.

Note that this creation process needs two photons and one neutrino to be accomplished. One question arises: where the little neutrino comes from?

It is proposed that there exist an abundance of neutrinos with negligible mass and negligible Kinetic Energy everywhere in the Universe. It's one of the basic particles of the Universe like electrons and photons and they are more concentrated around every material object in the Universe attracted by the Gravitational Fields.

NOTE:

Two photons, each with enough energy $E_{TOTAL} = m_f c^2 = m_e c^2$, colliding with an atom and interacting with a neutrino can also produce a pair. The three particles must have same electrical orientation. The Kinetic Energies of the photons are transferred to Electric and Magnetic energies in the neutrino's rings and a similar intermediate structure of six rings as above takes place. An electron and positron pair would also be created this way. This would be the way creation processes happen in experiments where high energy photons collide with atoms.

NOTE:

The described processes verify the Principle of Conservation of the Angular Momentum considering all the rings. They could not conserve the currently called "spin" of the particles which is not a direct measure of the Angular Momentum (see Section 4.10).

5.3   PHOTONS' ABSORPTION AND EMISSION

The emission and absorption spectrums of materials are discrete. Each atom and molecule has it characteristic spectrum in the type of photons emitted or absorbed. They emit and absorb a discrete amount of types of photons what may leave us to consider a discrete variation in the value γ for the photon. But if we consider all the materials possible in nature the general characteristic spectrum seems to be continuous so, we are leaved to conclude that there is a continuous possibility of variation of the value γ of the photons.

It is well known that atoms have energy levels and that there are a discrete number of possibilities for these levels. The atom absorbs from the photon or emits to the photon the discrete quantity of energy that results of the change of the energy of the atom from one level to another.

What is new within the theories presented here is what does energy level means for an atom. There's a new possibility for the electron rather than orbiting around the nucleus of the Rutherford-Bohr atom:

It is proposed that the electrons occupy fixed positions apart from the nucleus at a distance determined by the equilibrium of the electric attraction and the magnetic repulsion between each electron and the associated proton in the nucleus.

It must be considered that protons and electrons are particles constituted by rings of current. They interact with their Electric and Magnetic Fields and can reach equilibrium states. From the point of view of the electron, it "sees" a positive charge and also a Magnetic Field from the nucleus and the electron finds some positions where equilibrium states can be accomplished. Their behavior is similar to that of rings of opposite charges as described in Section 3.4 - Case A.

These positions depend on the possible variation of the γ value of the protons and the structure of the atom. The γ value of the electrons is constant.

It is proposed that the possibility of variation of the value γ of the photon is continuous and that the actual quantum behavior of the photons depends on a discrete possibility for the variation of the value γ of the protons in the atomic geometrical structure.

The protons and neutrons with a γ value that has small discrete variations accommodate in a geometrical structure in the nucleus. The current of the rings of each proton or neutron in the nucleus can have then small variations to accomplish an equilibrium state in the nucleus' structure.

Each electron in the atom also reaches an equilibrium state with their associated protons. These equilibrium states have an associated energy stored as the electromagnetic Potential Energy P_{EM} between the proton and the electron which can have small variations as the protons' γ value has small variations.

The quantum discrete possibilities for the energy of the emitted or absorbed photon determine that the possibilities for the different equilibrium states are discrete.

There must be a relationship between the variations in γ and the geometry of atoms with the quantum numbers of Quantum Physics. This needs further research.

I) Photons' absorption

In a photons' absorption, a traveling photon reaches an atom and interacts with a pair of associated proton and electron.

The photon delivers its Mass Energy E_m and its Kinetic Energy E_k to the atom that absorbs it in the form of an increment ΔE_m in the Mass Energy of the associated proton and a correspondent increment ΔP_{EM} in the Electro-Magnetic Potential Energy with the associated electron.

The proton and the electron reach another configuration with a higher level of energy. The total amount of energy delivered by the photon depends on the difference of energy in the quantum levels before and after the interaction.

It is proposed that the photon that loses its energy becomes a neutrino with negligible mass and Kinetic Energy.

The increments of energies of the atom are:

$\Delta E_m + \Delta P_{EM} = E_{TOTAL}$

where: ΔE_m is the increment in the Mass Energy of the proton and the electron

ΔP_{EM} is the increment in the P_{EM} of the proton and the electron

E_{TOTAL} is the total energy of the photon

II) Photons' emission

The photons' emission is a symmetric phenomenon of the photons' absorption. It must be assumed that normally there's a high amount of neutrinos around the atoms of matter in nature.

It is proposed that in a photons' emission a pair of associated proton and electron in a high energy level interacts with a neutrino with negligible mass and Kinetic Energy which becomes a photon.

In the interaction the proton and the electron jump to another configuration with a lower level of energy, the neutrino absorbs the variation ΔE_m in the Mass Energy of the proton and the variation ΔP_{EM} in the Electro-Magnetic Potential Energy that exist with the associated electron.

The neutrino absorbs all that energy in the form of its own Mass Energy E_m and Kinetic Energy (incrementing its velocity) until it reaches the final velocity ς becoming a photon. The final photon has always part of its energy in the form of Mass Energy and the other part in the form of its Kinetic Energy:

$$E_m = \tfrac{1}{2}m_f c^2$$

$$E_k = \tfrac{1}{2}m_f \varsigma^2$$

$$E_{TOTAL} = E_m + E_k = \tfrac{1}{2}m_f c^2 + \tfrac{1}{2}m_f \varsigma^2 = E_f$$

$$\varsigma \approx c \ \ then \ E_f \approx m_f c^2$$

The type of the photon created is determined by its mass' m_f value which just depends on the final total E_f energy acquired. Its length λ is determined by the De Broglie Law $\lambda = h/m\varsigma$ and its frequency v by the relation $\lambda v = \varsigma$ or the equation $E_f = h(v_0 + v)/2$ for the case when the source moves (Section 8.3).

A question arises: where does the neutrino of the interaction comes from? The photons' absorption is a source of neutrinos but also, as we mentioned above in the pair creation/annihilation phenomena, neutrinos are everywhere and in high amount.

5.4 QUANTIZED ENERGIES IN THE HYDROGEN ATOM

We know from Experimental Physics that quantum levels are expected in the emission and absorption of energies in the Hydrogen atom. We are going to see here how these conditions can be translated to quantum conditions in our model of the atom where electrons occupy fixed positions.

We know from Experimental Quantum Physics that the energy level in the Hydrogen atom is quantified: $E = E_n$

E_n jumps between quantum values:

$E_n = (1/n^2)E_1$

where $E_1 \approx 2.2 \times 10^{-18} J \approx 13.6\ eV$

and $n = 1, 2, 3...$

Considering the source at rest the energy of a photon verifies:

$E = hv$

Then, the frequencies of the photons corresponding to the maximal emissions or absorptions possible in those levels are:

$v = (1/n^2)E_1/h$

v assumes quantum values:

v is proportional to $1/n^2$

$v = v_1/n^2$

where $v_1 = E_1/h$

The λ is then:

$\lambda = c/v = n^2 c/v_1$

λ assumes quantum values:

λ is proportional to n^2

$\lambda = n^2\lambda_1$

where $\lambda_1 = c/v_1$

The photons verify the De Broglie formula so:

$m_f = h/\lambda c$

Then:

$m_f = (1/n^2)h/\lambda_1 c$

The mass of the photons is also quantified:

m_f is proportional to $1/n^2$

$m_f = m_{f1}/n^2$

where $m_{f1} = h/\lambda_1 c$

We will consider now the expression for the mass of photons that was delivered in Section 3.4 - Case A:

$m_f^2 \approx Kh^2/4A\gamma^2$

$\gamma^2 \approx Kh^2/4Am_f^2$

Then:

$\gamma \approx (K/4A)^{1/2}(h/m_f)$

$\gamma \approx n^2(K/4A)^{1/2}(h/m_{f1})$

Finally we found that the value γ must be also quantified:

γ is proportional to n^2

$$\boxed{\gamma = n^2\gamma_1}$$

where

$\gamma_1 \approx (K/4A)^{1/2}(h/m_{f1})$

$n = 1, 2, 3\ldots$

Each γ determines a different curve for the Repulsive Magnetic Force inside the emitted photons what at the end determines a different equilibrium point for the photons.

This means that each γ determines a different λ for the emitted photons as we can see in the next illustrative graphic.

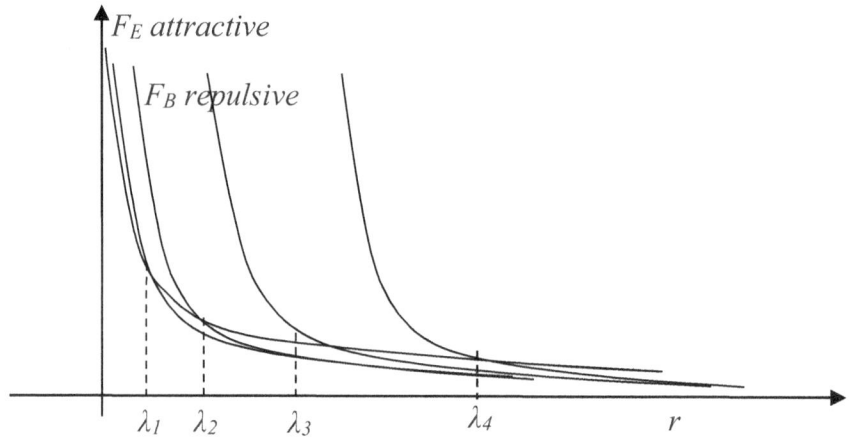

Since the electron's rings seem to have a constant γ value, the proton's rings must then have a γ value that varies discretely. The set of energies of the emitted or absorbed photons is given by:

$$E_n = E_1/n^2$$

The variation of the energy in the proton is then:

$$E_P = E_{P0} \pm E_n = E_{P0} \pm E_1/n^2$$

As the relation between energy and mass is $E = mc^2/2$ follows:

$$m_P = m_{P0} \pm m_n = m_{P0} \pm m_1/n^2$$

and so:

$$1/\gamma_P = 1/\gamma_{P0} \pm 1/\gamma_n = 1/\gamma_{P0} \pm 1/n^2\gamma_1$$

The possible set of variation of the γ_P value of the proton is then given by:

$$\boxed{\gamma_P = n^2\gamma_{P0}\gamma_1/(n^2\gamma_1 \pm \gamma_{P0})}$$

This is the quantum condition for the γ value of the proton that determines the quantum levels for the emission/absorption energies in the Hydrogen atom.

$\underline{n}$ is the main quantum number of four in the complete Quantum Physics Theory developed. We found here a new way to explain it.

The other quantum numbers would have a proper interpretation within the concepts of the new theories presented here. This hasn't been done yet. Other ones must come, for instance, from the small discrete variations in the equilibrium states of protons and electrons in the different structures of all the different atoms.

5.5 "QUANTUM TUNNELING"

It has been verified that statistically some charged particles trespass potentials barriers that at a first approach they would not be able. Heisenberg's theory about the uncertainty of the state of a particle and Schrodinger's theory about its probability are currently taken to explain this phenomenon resulting in the "quantum tunneling" interpretation.

As stated at the end of Section 3.4, in this new theory, Schrodinger's and Heisenberg's theories do not apply and so the "quantum tunneling" concept has no sense.

Another interpretation is possible for the phenomenon of crossing potential barriers.

The real state of electrons in an atom cannot be perfectly determined due to the "thermal noise" phenomenon (well known in electronics). In normal conditions, at a given temperature, there are a constant average number of photons with diverse energies received and emitted by the atoms of any material in a dynamical equilibrium such that the average level of the total energy of the atoms are maintained constant. Each electron will statistically interact with photons reaching higher energy levels, going after to lower levels while emitting new photons, interacting again with other photons and successively changing its state. Eventually an electron can gain enough energy to get out of its atom but as its original state is unknown and the energy of the interacting photon is also unknown it will escape with an unknown Kinetic Energy. Some of these electrons will have enough energy to pass some potential barrier that the average electrons do not pass.

5.6   PHOTONS IN DENSE MEDIUMS

Light, composed by photons or trains of photons, has lower velocities in dense mediums.

The photons are continuously decelerated and reaccelerated again while interacting with the atoms in the medium and so their constituting elementary particles are not in equilibrium.

The Corrected De Broglie Law to be verified requires the particles to be in their inner equilibriums states as described in Section 3.4 – Case A and is not valid for the average velocity of the photons in dense mediums.

The masses m and the internal energies $Em = mc^2/2$ of the photons do not vary but their length r, their absolute velocity ς and their Kinetic Energy $E_k = m\varsigma^2/2$ vary strongly in their interactions with the atoms.

Section 8.5 gives a new interpretation of Fizeau's experiment about the velocity of propagation of light in dense mediums which is in accordance with the Emission Theory of light.

CHAPTER SIX

Here the Davisson-Germer experiment is analyzed. [5]

The experiment can validate a great part of the new theories.

6.1   DAVISSON-GERMER EXPERIMENT

Davisson and Germer realized an experiment where it can be observed the "wave like" behavior of the electrons. They could verify experimentally the De Broglie formula for the electrons.

The Davisson-Germer apparatus is a vacuum glass tube which has in its interior an accelerator of electrons, a known crystal structured substance as target and an electron detector.

Next is a simplified schema of the experiment:

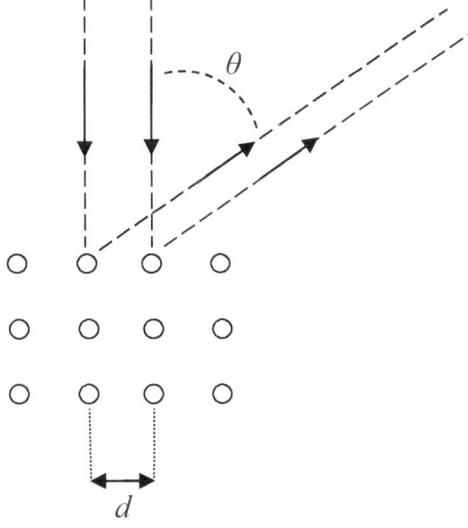

The accelerator is composed of two plates with terminals outside for connecting to an external voltage and with a hole in a plate to form a beam of electrons with certain velocity. The crystal target is where the electron beam collides and is diffracted. The electron detector can rotate around the target to detect in what directions the beam of electrons is diffracted.

The experiment showed that the electrons satisfied the relation for constructive interference:

$dsin\theta = n\lambda$

A beam of diffracted electrons is observed for the angle that satisfies the relation above for the De Broglie λ of the electrons:

$\lambda = h/mv$

Davisson and Germer used a formula that is delivered from the calculation of the velocity v of the electrons as a function of the known value of the voltage V between the accelerating plates that is assumed valid for slow velocity electrons as we see next.

The Kinetic Energy reached by the electrons equals the charge times the voltage through the plates:

$E_k = \frac{1}{2}mv^2 = eV$

and from the De Broglie formula we have:

$\lambda = h(2emV)^{-1/2}$

where $\underline{e}$ and $\underline{m}$ are the charge and the mass of the electron.

This value of λ is used in the experiment and verifies the constructive interference relation.

6.2   THE EXPERIMENT AT VERY SMALL VELOCITIES

We introduced at the beginning of this text a modification in the De Broglie formula.

The Corrected De Broglie formula is: $\lambda = h/mf(v_A)$

where v_A is the absolute velocity of the particle measured in a frame of reference at rest.

The theoretical conditions we found this function must satisfy are:

$f(v_A) \approx f(0) > 0$ for very small velocities

$f(v_A) \approx v_A$ for high velocities

$f(c) = c$

The next graphic shows an intuitive first approximation for the function:

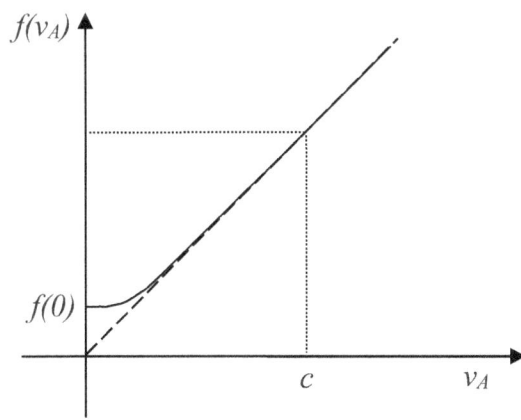

It is proposed that the function can be obtained experimentally through the Davisson-Germer experiment at very slow velocities.

6.3   THE EXPERIMENT AT HIGHER VELOCITIES

The original Davisson-Germer experiment is valid under the approximation of slow velocity electrons where Classical Physics applies. The following assumptions are assumed valid:

1) Kinetic Energy $= E_k = \frac{1}{2}mv^2$

2) Electrical Energy applied to the electrons $= qV$ (V: voltage)

Now we will consider the same experiment but at higher velocities. Important considerations must be taken on the assumptions above.

First, the new theories predict the mass of the electrons to be constant under velocity variations then, even at high velocities, the classical formula for the Kinetic Energy remains valid:

$E_k = \frac{1}{2}mv^2$ (m constant)

Second, with the new theories the second assumption is not valid.

It must be considered that the new Electric Force proposed is affected by the factor $s = (1 - v^2/c^2)^{1/2}$ which is velocity dependent.

It can be calculated that the Electric Energy applied to the electrons (which by definition is equal to the work done by the Electric Force) is:

$W_E = qU(1 - qU/(2mc^2))$

Where U is the real Electric Potential between the accelerator plates which is related to the measured voltage V as follows:

In Section 2.3-III it is derived the equation that relates the real Electric Potential U present in the linear electrical accelerator of Bertozzi's experiment and the final velocity v acquired by the electrons in accordance with the new definition of the Electric Force. The resultant equation is:

$$qU = (1 - s)mc^2 \qquad \text{where } s = (1 - v^2/c^2)^{1/2}$$

The equation is different than that considered by Bertozzi based on the measurement of a voltmeter and the relativistic Kinetic Energy:

$$qV = (1/s - 1)mc^2$$

It is pointed out there that the new theory would demand some important review in the functioning of the voltmeters and in the way their measurements are applied. It is argued that the voltmeter measures a "voltage" V which actually is related to the real Electric Potential U by the relation $V = U/s$. Substituting values is easy to observe that in this case both equations are equivalent.

It is considered that the equation of Bertozzi's experiment, which is used in electrical accelerators in general, works because of an extraordinary physical coincidence.

It must also be noted here that as pointed out in Section 4.6 common electrical accelerators do not guarantee the emission of individual electrons at a time. Actually, bursts of parallel trains of electrons would be emitted at a time.

6.4   AGREEMENT WITH THE EXPERIMENTS

As pointed out in the previous section the new theory agrees with the equation:

$qV = (1/s - 1)mc^2$

This is the considered equation in the Davisson-Germer experiment at high velocities with Relativity Theory.

In this new theory the mass does not vary with velocity as in Relativity Theory.

If right, this theory would have a better agreement with the experiments than Relativity has.

CHAPTER SEVEN

7.1 ELECTROMAGNETIC WAVES DO NOT EXIST

The deduction of the existence of "electromagnetic waves" from Maxwell Equations is wrong because of a missing step.

Once the planar waves are deduced as possible solutions to the set of four equations from Maxwell Equations it is absolutely necessary to ask: *Which source for the Electric and Magnetic Fields can generate those possible fields?* If not, if no source is related to the fields, we will be leaved to admit that infinite waves exist in the space of all frequencies, intensities and in any direction.

The solution for the Electric and Magnetic Fields is an infinite plane with the same (constant) field, parallel to the plane, in the entire plane. Sources for them are just impossible. There's no source of field possible to generate that kind of fields. Even an infinite series of those solutions would have no possible source for their fields.

Then, it can be stated that "electromagnetic waves" cannot exist.

As was presented in Section 1.1 – item III-e, the Electric and Magnetic Fields are assumed instantaneous. There is no delay in the action of the forces whatever the distance can be.

The electromagnetic wave signal transmission is the unique phenomenon that seems to prove that the Electric and Magnetic Fields propagate at the c finite velocity.

The experimental demonstration of the existence of electromagnetic waves was carried by Hertz. In the next sections it is presented a new interpretation of his experiments.

It will be shown that actually signal transmission is carried by photons.

7.2 HERTZ EXPERIMENTS

First we must note that the initial experiments of Hertz were about Electromagnetic induction. A vibrator produced high currents in an inductor made by rings. And an induced current was observed in a ring separated at some distance.

The electrons have mass and it will take time to accelerate them and although the Electric and Magnetic Fields are instantaneous, the phenomenon is not because is limited by the ς velocity at which an electron can theoretically be accelerated.

After Maxwell presented his equations electromagnetic waves were theoretically expected. Hertz performed an experiment that seemed to prove the existence of them.

The experiment consists of a radiator driven by an oscillator and a reflecting metal sheet.

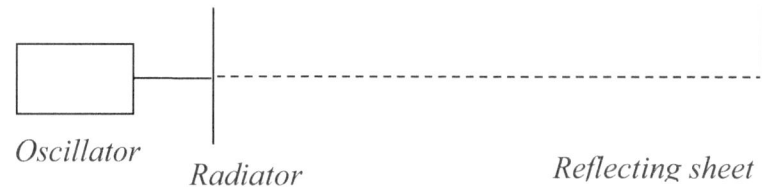

Oscillator

Radiator

Reflecting sheet

Moving the oscillator Hertz detected in some situations successions of maximums and zeros of the signal between the radiator and the reflector. These were the maximums and nodes of a procured standing wave. He measured the distance between the nodes and determined the "wave-length". With the frequency of the oscillator he could determine the velocity of the assumed traveling waves by the relation $\lambda v = c$.

This proved the existence of the electromagnetic waves.

It will be presented here a new interpretation of the phenomena.

It is proposed that in Hertz experiment the radiator actually emitted photons and what Hertz detected was not an electric or magnetic signal but photons' intensity.

Of course, the velocity of the traveling signal should be that of the photons velocity that is c.

The maximums and nodes detected by Hertz should correspond to maximums and the average intensity of photons respectively in that position. Photons are emitted and reflected by the metal sheet and what happens in each of the middle points is not interference but the addition of the intensity (this means quantity) of transmitted and reflected photons.

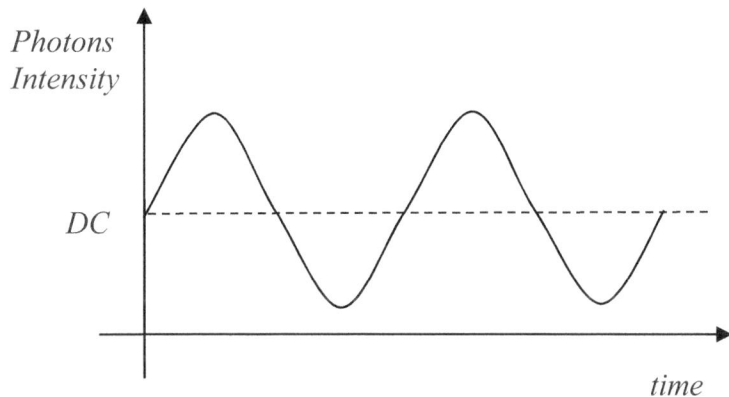

To interpret the phenomena in the right manner we must consider that Hertz used antennas to radiate and detect electromagnetic fields and it is proposed here that antennas emit or absorb photons.

The phenomenon is explained with more details in the next section.

7.3   COMMUNICATING WITH PHOTONS

All kinds of radiation like radio, TV, microwaves, infrared, visible light, ultraviolet, high energy rays and any kind of heat are all "electromagnetic radiation".

But what must be understood by "electromagnetic radiation" is radiation of photons ("electromagnetic particles").

It is proposed that what any antenna emits or absorb are photons. What actually happen in communications antennas are the photons' emission or absorption phenomena.

The electric circuit in a transmitter continuously extract and return back electrons to atoms of the metal structure of the antenna in a process where they emit some kind of photons with an intensity proportional to the intensity of the signal being transmitted.

At a receiver antenna photoelectric effect takes place absorbing those photons and liberating electrons of the antenna to be attracted by the receiver circuit generating a received signal.

The frequency of the emitted photons is different from the frequency of the communication signal. What is modulated is the intensity of the radiated photons. The kind of photons emitted or absorbed is what can produce photons' emission or absorption in the antennas.

There must be some "DCP" positive average level on the intensity of the photons being emitted. There cannot be negative values for the intensity of photons. The periodic signals of a typical communication modulate the intensity of the photons from zero to a maximum. This implies the transmitters must provide only positive values for the signal in the antenna and so the signal must have a correspondent "DCS" level.

The next graphics illustrate the phenomena for both the photons intensity and the respective signal intensity at the antenna terminal.

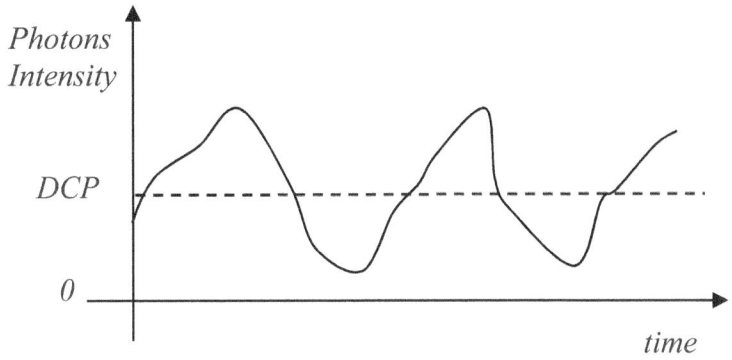

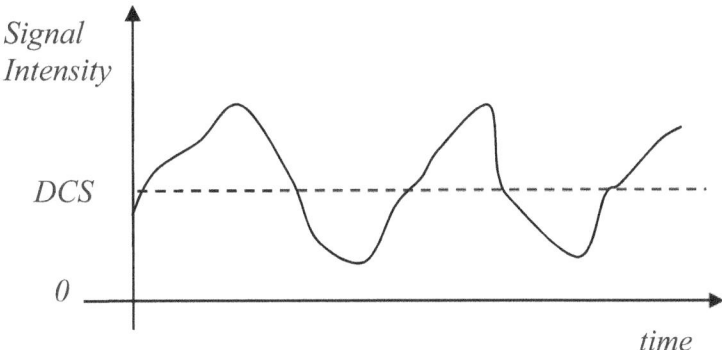

If we consider a system where reflected photons exist, it can be demonstrated that all the properties of the waves are conserved. The intensity of the reflected photons is added to the direct photons and another wave is achieved. Especially a standing wave pattern can be produced by the photons.

Now, if we consider the possibility of communications by photons a fast question arises:

QUESTION:

A question that may surge if we consider the possibility of communication by photons is why a typical linear antenna needs to have a length equal to $\lambda/4$.

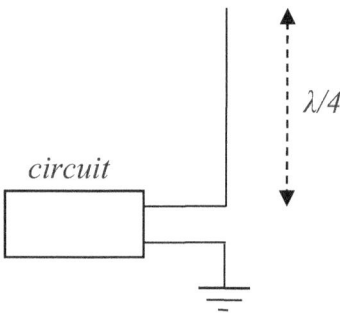

To answer this question we must treat the problem of the velocity that electrons travels in an antenna.

With the new theories is possible for electrons to travel at velocity c and it is proposed to consider the possibility that the electrons in antennas do travel at velocity c. It must be considered that the electrons travel through the molecular structure of the metallic antennas and can be accelerated by positive protons at very small atomic distances where the Electrical and Magnetic Forces can be very strong and the electrons can reach velocity c.

As the electrons moves from an antenna to the circuit and inversely, the signal at the input terminal of the circuit varies.

Let suppose that initially the signal is at the DCS value but photons are being received by a receiver antenna. Then, it gets more positively charged and the terminal extreme has a more positive signal.

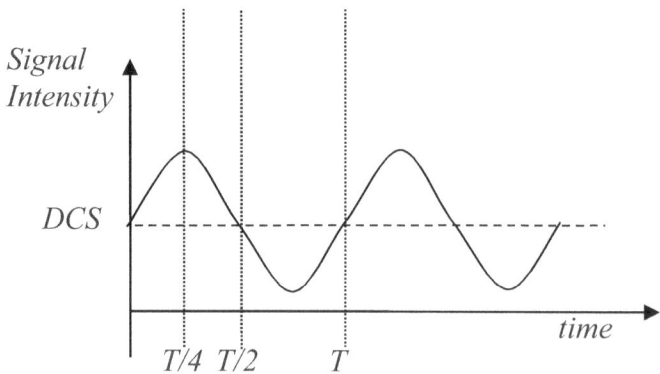

The signal in the terminals is initially DCS and it will take a quarter of the period T of the signal to reach its maximum positive value. In that interval of time the electrons moved from the antenna to the circuit and the last ones have traveled a distance equal to $\lambda/4$ if their velocity is c.

In the next $T/4$ interval the photons intensity decrements and the antenna gets less positive while it gain again the electrons from the circuit.

A similar reasoning can be applied in the next $T/2$ interval where first electrons move from the circuit to the antenna returning back after.

An antenna of length $\lambda/4$ corresponds to the maximum number of electrons that can be taken out of the circuit in a quarter of the period of the signal.
This will be the case of a maximum signal detected by the circuit.

If the antenna has a length larger than $\lambda/4$ it will be capable to detect signals of other frequencies and more noise is added to the receiver circuit. Then, the optimum length of the arm of the antenna will be $\lambda/4$.

A symmetrical reasoning applies in a transmitter.

This is the same condition reached by the electromagnetic wave theory and the experience.

CHAPTER EIGHT

Here we present the Emission Theory and some results.

8.1 THE EMISSION THEORY

The Emission Theory is consequence of applying directly Newton's dynamics to light as particles with mass emitted by material objects. It proposes that the velocity of light is the constant c plus the velocity of the source of the light.

In vectors form:

$$\varsigma = c + u$$

Without any convincing proof and with the rise of the Relativity Theory the Emission Theory was forgotten.

No variations in the speed of light have been experimentally detected until today. Several reasons can cause this invariance.

One reason is that in nature the possible sources move too slowly to be detected.

Other one is that, as Emission Theory states, every photon acquires a new source velocity component (u) when they interacts with atoms (like those of glasses, mirrors, those of the atmosphere and even those of the interstellar gases in the vast space) so the original source velocity component is lost.

Yet another one surges now due to the train structure of the light's rays of the stars that reaches Earth. For instance, in the double stars' emission phenomena the emitted photons would just form trains of photons with different lengths and spacing in one half of the cycle than on the other half. There would be not any strange effect in the resultant rays.

It must be pointed out that the Emission Theory verifies Michelson-Morley experiment as described in the next section.

8.2 MICHELSON-MORLEY EXPERIMENT

In that years many physicists believed that the light should move in a particular medium in the space called the aether.

The experiment was conceived to detect any variation of light velocity in the possible moving aether. A negative result was obtained and was proved that the aether does not exist.

We will see here how the Emission Theory is consistent with the Michelson-Morley experiment.

The Michelson-Morley apparatus consist of a source of light and a special disposition of mirrors as shown in the figure.

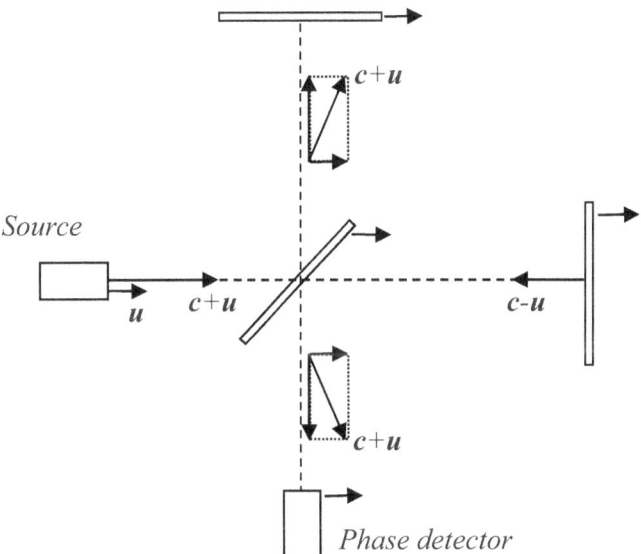

The central mirror is only partial reflective so that part of the beam is reflected but part of it passes through. The other mirrors are totally reflective. Then, a beam of light is divided into two beams by the central mirror. Each one is reflected in a different mirror and they both come back to the central one. The beams are partially reflected again and they finally join and a composed beam reaches the interference detector. The total apparatus moves with Earth. Then the source, the mirrors and the detectors all have a velocity u. If a difference in the relative velocity of the beams respect to the mirrors exists in one path a phase difference would be observed in the phase detector.

The Emission theory states that the light has a velocity vector $ç$ that is the vector addition of the constant velocity vector c in the direction of the emitted beam and the source velocity vector u.

$$ç = c + u$$

When a beam of light is reflected by or passes through a mirror it interacts with in such a way that the mirrors act as a new source for the beams. This means that the light wins a new velocity vector each time a beam interact with a mirror as shown in the figure.

Respect to the mirrors all of the beams have the same relative velocity.

In other way, calculations of the absolute velocities can be made to show that the two final beams reach the detector in phase. Then, no phase difference is theoretically expected.

8.3 WHEN THE SOURCE MOVES

It is interesting to calculate the parameters length and frequency of a beam of photons when its source moves with velocity u.

When the source moves the beam of light has velocity $ç$:

$$ç = c + u$$

The De Broglie formula states that the light's length is:

$$\lambda = h/mç$$
$$\lambda = h/(m(c + u)) = h/(mc(1 + u/c)) = \lambda_0/(1 + u/c)$$

$$\boxed{\lambda = \lambda_0/(1 + u/c)}$$

The light's frequency is:

$$v = ç/\lambda = c(1 + u/c)/(\lambda_0/(1 + u/c)) = v_0(1 + u/c)^2$$

$$\boxed{v = v_0(1 + u/c)^2}$$

The energy of a photon is: $E = Em + Ek$

$$E = mc^2/2 + mç^2/2 = mc^2/2 + mc^2(1 + u/c)^2 = h\,v_0/2 + h\,v_0(1 + u/c)^2/2$$

$$\boxed{E = h(v_0 + v)/2}$$

8.4 DOPPLER AND TRANSVERSAL DOPPLER EFFECTS

DOPPLER EFFECT:

By definition the Doppler Effect is the variation in the light's frequency when the source or the observer moves for a beam emitted parallel to the movement.

We have already calculated in the previous section the light's length and frequency when the source moves with velocity u:

$$\lambda = \lambda_0/(1 + u/c), \quad v = v_0(1 + u/c)^2$$

$$\boxed{\textit{The Doppler effect for a moving source: } v = v_0(1 + u/c)^2}$$

The beam can have the same or the opposite direction of the velocity **u** of the source then u will be positive or negative respectively.

We observe now that λ doesn't change when the observer moves then:

$\lambda = \lambda_0, \ \upsilon = \upsilon_0(1 + u/c)$

The Doppler Effect for a moving observer: $\upsilon = \upsilon_0(1 + u/c)$

Note that the color of light as seen by our eyes could actually be directly related to its energy and so to υ.

TRANSVERSAL DOPPLER EFFECT:

The Transversal Doppler Effect is by definition the variation in the light's frequency when the source moves for a beam detected in a perpendicular direction of the movement.

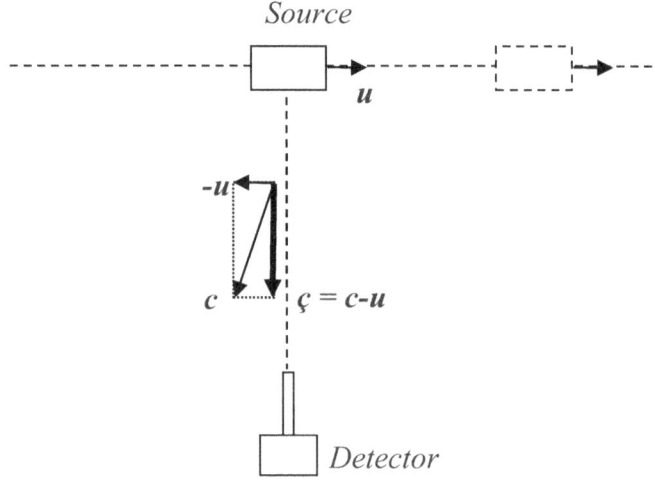

We must carefully observe that to get a beam that travels in a direction perpendicular to the trajectory of the source it must be emitted a beam backwards such that the resultant absolute velocity is:

$$\varsigma = c - u$$

The absolute value of the velocity is then:

$$\varsigma = (c^2 - u^2)^{1/2} = c(1 - u^2/c^2)^{1/2}$$

Then, the light's length and frequency are:

$$\lambda = h/m\varsigma = h/(mc(1 - u^2/c^2)^{1/2})$$

$$\upsilon = \varsigma/\lambda$$

$$\upsilon = \upsilon_0(1 - u^2/c^2)^{1/2}$$

$$\boxed{\upsilon = \upsilon_0(1 - u^2/c^2)^{1/2} \text{ for the Transversal Doppler Effect}}$$

Then, υ is reduced and we have the same Transversal Doppler Effect as that proposed by the Relativity Theory.

8.5 FIZEAU EXPERIMENT

Light has in a dense medium a relative average velocity w slower than its velocity in vacuum c due to the successive interactions of the photons with the atoms. The photons are slowed down in a little interval of time and in a small distance near the atoms. Between the atoms they recover its vacuum velocity.

Fizeau's experiment measures the light velocity in a liquid medium that travels at velocity u. Fizeau determined that if the velocity of light in the medium at rest is w_0 then the average velocity of light $ç_A$ relative to a frame at rest is:

$$ç_A = w_0 + u(1 - w_0^2/c^2)$$

where c is the light velocity in vacuum.

It is proposed here that two phenomena should be taken into account in the experiment:

a) *The medium is moving with velocity u then u should appear as a source component of the final absolute velocity.*

b) *The velocity of light relative to the medium is dependent on the velocity of the medium: w = w(u)*

With the Emission Theory the absolute velocity is:

$$ç_A = u + w(u)$$

Then, Fizeau experiment gives us the function of the relation of w with u. If we take the Fizeau expression and reorganize terms we have:

$$ç_A = u + w_0 (1 - uw_0/c^2)$$

Then:

$$\boxed{w = w(u) = w_0 (1 - uw_0/c^2)}$$

NOTE:

It's important to note that the analogy between the equivalence between the formulas of the addition of velocities in Relativity Theory and that obtained by Fizeau in his experiment is valid only under a questionable approximation.

The formula of Relativity for the experiment described above gives us:

$$\varsigma_A = (u + w_0)/(1 + u\,w_0/c^2)$$

8.6 SAGNAC EFFECT

Sagnac effect is well presented in many sites on the web. It is sustained that it disagrees with the "ballistic" (Emission Theory) concept of light for which no interference pattern would appear as it does in practice.

The problem is that it is assumed that reflections of a beam of light happen instantaneously while within the new proposed theory they take some interval of time which depends on the velocity of the beam. This introduces velocity dependent phase shifts in the reflected beams and so velocity dependent interference patterns in the Sagnac experiment are expected.

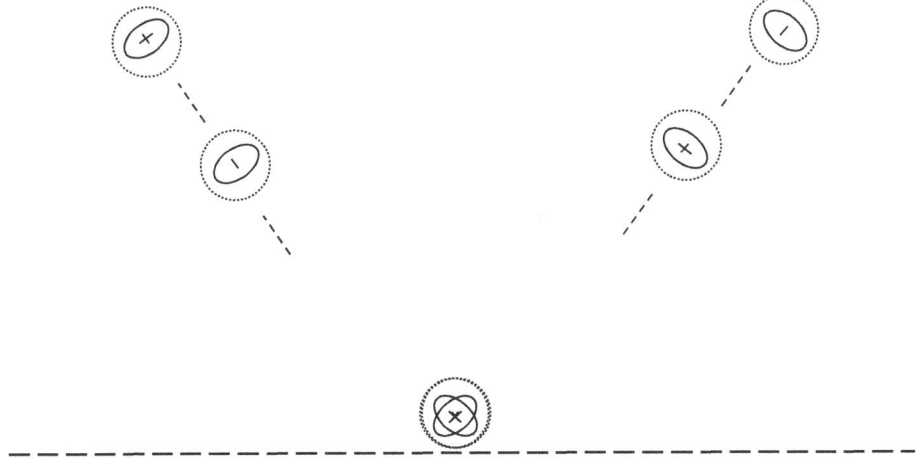

In Section 4.1 it is described the structure of a photon as composed by two elementary particles (a positrin and a negatrin) traveling together at half the *De Broglie distance* $\lambda = h/m\varsigma$ with velocity $\varsigma = c + u$ (vectorial addition) where u is the absolute velocity of the source.

Reflections happen through absorption and re-emission of photons.

In a reflection of a photon with light distance $\lambda_1/2$ where $\lambda_1 = h/m\varsigma_1$ and *velocity* ς_1, it takes some time T_1 for the photon to be absorbed. A new photon is created which will have a new light distance $\lambda_2/2$ and ς_2 with $\varsigma_2 = c + u_2$ where u_2 is the absolute velocity of the mirror and $\lambda_2 = h/m\varsigma_2$. It will take some time $T2$ until the photon is completely released from the surface. The total reflection process took an interval of time $T = T_1 + T_2$ to take place and so a "phase-shift" in the reflected beam has been produced. Note that the value of the "phase-shift" depends on the absolute velocity u_2 of the mirror and so a velocity dependent "phase-shift" would be produced in agreement with the experiment.

In a Sagnac apparatus the phases of the two beams produced in opposite directions will be affected differently by the mirrors due to their absolute velocity and so when arriving at the detector a phase difference on the two beams will be present which produces the observed interference pattern.

Note that when the velocity of the reflected photons is higher than the incoming ones they would become a little shorter and more spaced. When the reflected photons have a lower velocity they would become a little larger and less spaced.

APPENDIX A

DEMONSTRATIONS OF RELATIVITY THEORY WRONG

A) The real equation of force is $F = ma$

Today's Physics is stating that the Equation of Force is $F = dp/dt$.

We will analyze the equation of motion of rockets to see that the real Equation of Force is:

$$\boxed{F = ma}$$

A rocket has variable mass in its trajectory and it's important to see its motion equation.

Let m be its variable mass at any instant in its movement composed by the mass of the rocket plus the mass of its contained fuel.

I have made a search in the internet about rocket motion equations and all the sites agree in the equation: [6]

$F = m(dv/dt) = -v_e(dm/dt)$ where v_e is the speed of the expelled fuel relative to the rocket.

They all agree that the force acting on the rocket is due to the expelled mass and is $F = -v_e(dm/dt)$ and that the equation of motion is $F = m(dv/dt) = ma$.

I assume the equation has been completely verified experimentally with enough precision from a long time ago.

It is evident that in the development it is used covertly the equation:

$F = ma$

 for the force and not:

$F = dp/dt$

By definition $p = mv$ and $dp/dt = m(dv/dt) + v(dm/dt)$.

They derive the rocket's equation of motion based on the principle of conservation of momentum but considering the momentum of the rocket with the totality of the fuel (the contained plus the expelled fuel) at any moment and stating $dp/dt = 0$. After that they derive the equation of motion of the rocket as: $m(dv/dt) = -v_e(dm/dt)$ and specifically say that the force on the rocket is:

$F = m(dv/dt) = -v_e(dm/dt)$

$m(dv/dt) = ma$ then it is clear that what is finally applied to the rocket to determine its movement is the equation $F = ma$ and not $F = dp/dt$.

This indicates that today's Physics is wrong stating the Equation of Force as $F = dp/dt$.

The right Equation of Force is $\boldsymbol{F} = \boldsymbol{ma}$ even when mass varies.

Note that the natural derivation of the famous equation $E = mc^2$ by Relativity Theory has no sense since it is based in the wrong relation $F = dp/dt$.

Relativity Theory becomes a wrong theory since it is based on a wrong law.

By definition:

$p = mv$

With partial derivatives:

$dp/dt = m(dv/dt) + v(dm/dt)$

Now as:

$F = m(dv/dt)$

Then, the valid Equation of Momentum and Force is

$$dp/dt = F + v(dm/dt)$$

Then, the principle of conservation of the momentum $p = mv$ must determine that $dp/dt = 0$ when no forces are applied and when there's no variation on the considered mass.

It can be observed that this principle can be applied to the rocket as was applied in the cited cases giving the same result. Considering the mass m' of the composed body of the rocket and the total fuel (the contained plus the expelled fuel) which does not vary:

$F' = 0$ and $dm'/dt = 0$

Then, the thrust equation can be derived (as below):

$m(dv/dt) = -v_e(dm/dt)$

where m is the mass of the rocket with its contained fuel.

Finally:

$F = ma = m(dv/dt) = -v_e(dm/dt)$ is the force exerted on the rocket.

$F = -v_e(dm/dt)$

Rocket thrust force derivation

The thrust equation of the rocket is derived here considering the approximation that the mass of the expelled fuel of the rocket is negligible compared with the mass of the mass of the rocket plus the mass of its contained fuel:

Momentum and Force equations:

$\boldsymbol{p} = m\boldsymbol{v}$, $\boldsymbol{F} = m\boldsymbol{a} = md\boldsymbol{v}/dt$

$d\boldsymbol{p}/dt = md\boldsymbol{v}/dt + \boldsymbol{v}dm/dt = \boldsymbol{F} + \boldsymbol{v}dm/dt$

Considerations:

a) masses equations:

m = mass of the rocket with its contained fuel

m_e = mass of the expelled fuel

m' = m + me = *constant* = total mass of the system rocket with total fuel

$dm'/dt = dm/dt + dm_e/dt = 0$

Then: $dm_e/dt = -dm/dt$

b) velocities equations (one dimension):

v = absolute velocity of the rocket

v_e = velocity of the expelled fuel in relation to the rocket assumed constant

u = absolute velocity of the expelled fuel

$v_e = v - u = constant$

$dv_e/dt = dv/dt - du//dt = 0$

Then: $du/dt = dv/dt$

Derivation:

Total momentum of the system rocket with total fuel: $p' = mv + m_eu$

$dp'/dt = d(mv + m_eu)/dt = mdv/dt + v/dm/dt + m_edu/dt + u/dm_e/dt$

Considering: $du/dt = dv/dt$ and $dm_e/dt = -dm/dt$

$dp'/dt = (m + m_e)dv/dt + (v-u)dm/dt$

As $v - u = v_e$ and considering $m_e << m$ ($m + m_e \approx m$) then:

$dp'/dt \approx mdv/dt + v_edm/dt$ and as $dp'/dt = 0$ then:

$mdv/dt \approx -v_edm/dt$

Finally $F = ma = mdv/dt \approx -v_edm/dt$ under the approximation $m_e << m$

Then, the rocket thrust force is: $F \approx -v_edm/dt$

B) Pendulums and gyroscopes

Another consideration against Relativity is about the Foucault pendulum and the gyroscope behavior that is related to the conservation of the angular momentum of the bodies. If there is no special Rest Frame of Reference then we may wonder: related to what frame of reference are the directions determined by these apparatus fixed?

Fixed directions in space can be determined by pendulums and gyroscopes. Rest Frames at rest in the Universe is more difficult to determine since their center must be at rest, without any movement.

The center of the Universe would be at rest and would be a right origin for those frames (although difficult to determine).

Then, Rest Frames of Reference at rest, without movement, exist.

Objects at rest relative to Rest Frames are at absolute rest, they have no movement. Objects with not zero velocities on them have movement.

C) Absolute Magnetic Force

The Magnetic Force generated by a static Magnetic Field B over a charged particle moving at velocity v is defined by the Lorentz formula:

$$F_B = q.v \times B$$

The problem is which velocity must be considered.

The theory here agrees with Classical Physics and the v velocity that must be considered is the absolute velocity v_A relative to a frame at rest in the Universe. The Magnetic Force would be an Absolute Force.

We will consider an electromagnetic dynamics problem and study the classical and the relativistic predictions.

A beam of electrons is traveling in a straight line at some small velocity (generating a current which generates a Magnetic Field around) in relation of some frame at rest for the classical approach and just some frame for the relativistic approach.

Case 1:

The beam is traveling at velocity v_1 (generating a current q. v_1).

An electron is at rest in relation to the frame at some distance of the beam.

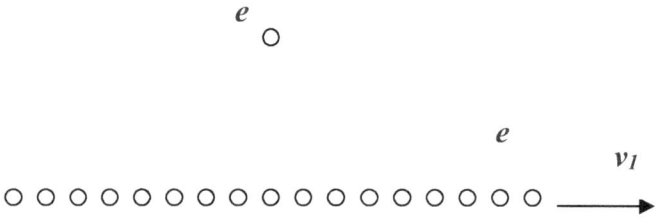

Case 2:

The electron now is moving parallel to the beam at some velocity w but the beam now is moving with a velocity $v_2 = v_1 + w$ incremented exactly by w (generating a current q.v_2).

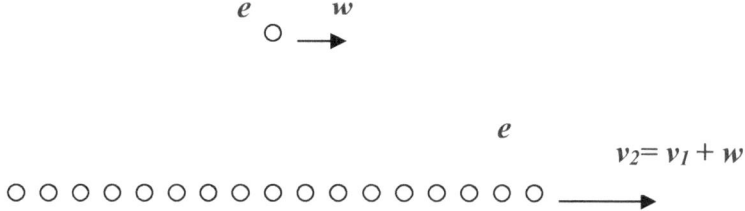

Classical Physics prediction is that in Case1 the Magnetic Force is $F_B = 0$ since $v_A = 0$ while in Case 2 it will be equal to $F_B = e.v \times B_2$ where e is the charge of the electron and B_2 the Magnetic Field generated by the current $i_2 = e. v_2$.

The total force that will act on the isolated electron is the addition of the Magnetic and the Electric Forces the beam generates and is different in each case. In the two cases the isolated electron will have a different trajectory in relation to the beam.

Classical Physics directly applies the Lorentz formula considering that in the two cases the beam and the electron have different absolute velocities relative to a frame at rest in the Universe and so they represent two different physical phenomena and the different trajectory of the movement is well justified.

In Relativity Theory there's no privileged Rest Frame of Reference so the two cases represent the same system or phenomenon just seen from different frames and the result should be the same. The relative movement between the isolated electron and the beam should be the same.

The two predictions would be contradictory and would show a case where Classical Physics and Relativity Theory would be not equivalent even considering slow velocities in disagreement with what is currently sustained.

It seems obvious that the classical approach would be the right one but may be an experimental verification of the phenomenon would be needed.

APPENDIX B

NEW INTERPRETATIONS FOR "RELATIVISTIC" PHENOMENA

A) A new interpretation of Mercury's orbital precession.

Classical Physics can explain only part of Mercury's orbital precession considering the effect of all the planets in the entire solar system.

A physical phenomenon that can explain the other component of the precession of Mercury's orbit differently from the relativistic interpretation is presented in this page. Only Classical Physics laws are applied.

The possibility that Sun's mass distribution could be symmetrical but not uniform must be considered as a possible cause of some of the Mercury's orbital precession.

The mass distribution of the Sun can have axial or planar symmetry through its axis of rotation, this way, and with some particular angular rotation and phase relative to Mercury's orbit the precession can be produced.

The distribution must be symmetrical since the gravity center of the Sun appears to be at the centre of the observable spherical shape of it.

Let consider an extreme case just as an illustrative case of not uniformly distribution. Let suppose the Sun has its mass divided in two halves concentrated in two points diametrically opposed in the apparent sphere:

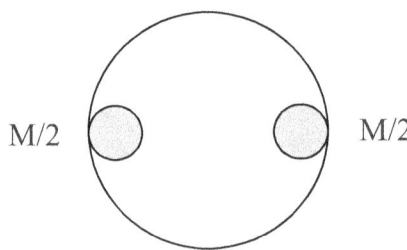

M/2   M/2

Let consider that Sun rotates such that the two concentrations of mass rotate in a plane with some direction that approximates to the plane of Mercury's orbit.

Let also consider that the period of rotation is about the half of the period of the orbit of the planet (Mercury's year) and that they are synchronized in a movement as illustrated in the following figures:

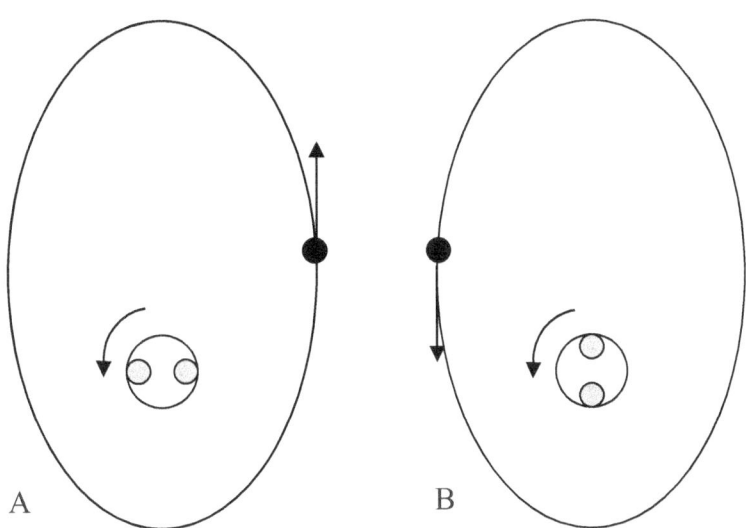

In the two halves of the orbit the planet will "see" the two concentrations of mass of the Sun in different manners. At the right half of the orbit the concentrations seems more aligned with the planet and in the left half they appear more transversal to it. These two cases determine two different gravitational forces acting in the two halves orbits.

It can be demonstrated by geometry that in the right half (figure A) the force is stronger than in the left half (figure B).

It is evident that this difference will introduce an asymmetry in the orbit. The asymmetry can be seen as a continuous displacement of the axis of an elliptical orbit (anti counter clockwise in the figures) at each cycle and that is the precession of the planet's orbit.

The effect is also present at sub-multiples of the period (multiples of the angular frequency).

Actually the period of the rotation of the Sun must be a little smaller than the half of Mercury's orbit period to present the same asymmetric pattern as the orbit changes its axis' direction.

The case presented is a very particular one of a hypothetical possibility of non-uniform mass distribution of the Sun well suited to explain how the precession can be generated. The real mass distribution of the Sun is not really known with enough precision.

One possibility to be taken into account is that the Sun could actually be composed by a not complete fusion of two smaller stars.

B) Bending of light by Gravitational Fields

In the new theories the photons have mass. For a source at rest:

$E = mc^2 = hv$

then:

$m = hv/c^2$

The problem of the bent of rays of the star's rays by the sun has been calculated for photons with the mass derived as above but only the half of the deflection measured experimentally was obtained.

The new theories introduce another phenomenon to be taken into account, is the fact that in general photons travel in arrays of trains of photons. The calculations must be made for trains of linked photons.

This will produce a different result in the calculations. This hasn't been done yet. They present some complexity that could be solved with computational methods.

C) About "time dilation"

Two main kinds of phenomena seem to prove the existence of time dilation:

1) Accelerated muons lifetime.

2) Atomic clocks traveling at "high" velocities.

Both phenomena are based on basic particles interactions.

The muons' elementary particles interactions towards the muons' "decay" and the natural photons' emissions of atoms in atomic clocks.

Those processes have now a new interpretation:

In the new theories the structures of all the basic particles and even the atoms and molecules are based on equilibrium states of the internal forces. The equilibrium states are velocity dependent mainly due to the $f(v)$ factor of the Corrected De Broglie relation.

It is predicted that "matter" structure is velocity dependent (not its mass which remains constant as pointed out in Section 3.4 – Case A).

The processes above are then expected to be velocity dependent and take different times to happen at different velocities.

In the new theories the forces are weaker as velocity increases and so the time for the processes to happen is expected to be larger in accordance with the experiments.

D) Summary of other interpretations for "relativistic" phenomena

Here is presented a summary of other experiments which have here in this text an alternative interpretation to that of Relativity Theory:

1) Kaufmann-Bucherer-Neumann experiments treated in Section 2.3.
2) The strong magnet experiment treated in Section 2.3.
3) Bertozzi experiment treated in Section 2.3.
4) Creation/annihilation related to equation $E=mc^2$ treated in Section 5.2.
5) Transversal Doppler Effect treated in Section 8.4.
6) Fizeau experiments treated in Section 8.5.
7) Sagnac effect treated in Section 8.6.
8) Davisson-Germer experiment treated in Chapter Six.

APPENDIX C

CORRECTIONS IN THE GRAVITATIONAL FIELD

A) Gravitational Field at different scales

Observations of the dynamics of spiral galaxies show an apparent invariance in the shape of the arms of the spiral what implies a constancy of the angular rotation in the stars of the arms. This dynamic does not agree with the classical prediction where stars more distant from the center would move slower.

The author believes in the possibility that the real Gravitational Field formula could be different. After all, Newton didn't have the information of the behavior of galaxies to be considered in his formulation of the Gravitational Field.

The classical approach could be an approximation valid at the scale of the solar planetary system but not at galactic scale.

If the Gravitational Field would be proportional to $1/r$ at galactic scale stars would have a constant angular rotation.

It is possible to find a right function that could make the Gravitational Field both approximately proportional to $1/r^2$ at the distances of a planetary system ("planetary scale") and approximately proportional to $1/r$ at the distances of galaxies' arms ("galactic scale") for example with two terms each one being the relevant one and the other with negligible effect in each case.

B) Reachability of the Field

The possibility that the Gravitational Force does not extend to infinite exists. A "reachability factor" could exist in the field's formula.

The factor would be a function of the distance "$f(r)$" which for some R_{max} would verify:

$f(r) > 0$ for $r < R_{max}$
$f(r) = 0$ for $r \geq R_{max}$

Note that the factor needs to be approximately "1" at relatively small distances where the classical prediction is valid and "0" at large distances larger than R_{max}.

Also note that this formulation determines continuity of order 1 in the fields but an unavoidable discontinuity at the derivative of some order necessary to have a finite limit.

The "reachability factor" could imply that gravity does not go farer than the region of the galaxies and that there is no gravitational attraction between well separated ones. One argument for this assumption is that all observed galaxies are finite in their extent what would be caused by a Gravitational Field restricted to a finite region in Space.

APPENDIX D

ABOUT HUBBLE RED-SHIFT

Physicist Hubble observed the light's spectrum of galaxies and found a red-shift obeying a pattern in which the red-shift varies progressively from far galaxies to near ones.

The interpretation of his observation was that galaxies are moving away in all directions and in an accelerated manner with nearer galaxies moving at higher velocities what leaved to the "Universe Expansion" theory on the history of the Universe and the "Big Bang" theory on its origin.

Here is suggested to consider the possibility that the variation in the spectrum of the galaxies could be produced because the kind of the photons emitted by the stars and so their length ("wavelength") could have changed through time.

Some parameter(s) in the Physics of the Universe, determining the kind of the photons emitted by the atoms in the processes of galaxies stars radiation, could have been varying through time in such a way that the observed red-shift in the spectrum is produced.

The new theories here give the precise structures for the photon, basic particles and the atoms and in principle the effect of the variation of some physical parameters in the emitted photons could be determined. The problem is that very sophisticated computational simulations of the emission of photons by some atoms, molecules and the physical processes involved in the photons' radiation by the stars seems to be needed.

NOTE:

New theories on the history and origins of the Universe would be needed.

FURTHER DEVELOPMENTS

SOME FURTHER DEVELOPMENTS THAT WOULD BE NEEDED BY THE NEW THEORY:

_ Davisson-Germer experiment must be performed at very slow velocities as proposed in Chapter Six.

_ A modified version of the Young double slit experiment must be performed providing a precise way of determining the exact number of particles that are present at each discrete emission event.

_ Bertozzi experiment must be reviewed for a new interpretation of the results as pointed out in Section 2.3.

_ As suggested in Section 4.1 a new calculation of the "bending of light" by the Sun considering light as trains of photons must be made.

_ The $f(v_A)$ function and the A constant introduced in Section 3.3 remains to be determined for the new theories.

_ A new interpretation for the atomic quantum numbers must surge (now exist the special value gamma).

_ The theoretical models of the experimentally detected particles in High Energy Physics need further development.

_ The theoretical determination of the Electromagnetic Potential Energy P_{EM} of the basic particles like the photon from the definition of the Electric and Magnetic Fields remains to be done.

CONCLUSION

Many propositions that could become new principles in Physics are presented in this text. They complete a new theory about light, the elementary particles and the basic fields and forces of nature.

The theory is successful in explaining all the following Physics subjects:

_ The "wave-like" behavior of the photons, electrons, protons and neutrons basic particles.

_ The double slit experiment.

_ The De Broglie formula although corrected and with a new physical meaning.

_ The $E = mc^2$ formula with a different physical meaning.

_ The "strong magnet", Kaufmann-Bucherer-Neumann and Bertozzi electrodynamics' experiments with electrons.

_ Why the magnetic force between magnets can be a function of exponent four in the distance.

_ The stable "spin" of the basic particles.

_ The chemical bonds of shared electrons between atoms in a molecule are well justified by the fixed positions of electrons in the atoms.

_ Davisson-Germer experiment.

_ Hertz experiments.

_ Michelson-Morley experiment.

_ Doppler and Transversal-Doppler effects.

_ Fizeau experiment.

_ Sagnac effect.

_ Light curvature.

_ Double stars' images.

_ Mercury's orbital precession.

_ Experimentally observed "time dilation" phenomena.

_ "Quantum Tunneling" phenomenon.

_ The quantum phenomena of pair's annihilation/creation, photons' absorption and emission effects and the quantization of energy levels in atoms.

_ "Subatomic" experimentally detected particles.

FINAL NOTE

The proposed new theory is consistent with Classical Physics, Photon's Physics, the Einstein $E=mc^2$ formula, Planck $E=h\upsilon$ formula and the De Broglie relation, although some corrections must be made.

It disagrees with Einstein's Relativity Theory, the called "Wave Mechanics Theory", the Electromagnetic Wave Theory, the Rutherford-Bohr model of the atom and today's subatomic "Standard Model" based on the Quarks Theory.

Some not good interpretations of some experiments and very bad unlucky coincidences have happened in the past and they make everyone think wrong.

May be the worst unhappy one is the De Broglie's proposition of a wave associated to matter that derived in the Schrödinger's equation and the Wave Mechanics Theory.

It was a very bad coincidence that the formula he proposed seemed to be confirmed by a later experiment and seemed to prove the theory while the formula is valid because of another physical phenomenon.

Something similar happens with Einstein's Relativity Theory.

If these didn't happen the history of Physics would have been totally different.

I have discovered those unlucky and unhappy errors and how to solve them, but I'm an Electrical Engineer not a Physicist. Is up to real physicists to take my work and develop a "New Physics". My work should be understood as a start-point. That's why I'm presenting the new theories here. It should be considered as I have solved the "engineering part" of the theories.

I have no more time, no more resources and no expertise to develop it further.

REFERENCES

BOOKS:

1) General Physics:

Fundamentos de Física

Frank J. Blatt

Tercera Edición

Prentice-Hall Hispanoamericana S.A.- 1991 (Spanish version)

2) Relativity Theory:

Sobre la Teoría de la Relatividad Especial y General

Albert Einstein

Debate – 1998 (Spanish version)

WEB LINKS:

3) Force between two magnets:

www.exo.net/~pauld/activities/magnetism/forcebetweenmagnets.html

4) Magnetic Field of a ring:

http://teacher.pas.rochester.edu/phy122/Lecture_Notes/Chapter30/chapter30.html#Heading1

5) Davisson-Germer experiment:

http://hyperphysics.phy-astr.gsu.edu/hbase/davger.html#c1

http://hyperphysics.phy-astr.gsu.edu/hbase/quantum/davger2.html#c1

6) Rocket dynamics:

http://www.braeunig.us/space/propuls.htm

AMAZON

www.amazon.com

`

www.ingramcontent.com/pod-product-compliance
Lightning Source LLC
Chambersburg PA
CBHW081301170526
45165CB00011B/3368